Colloids and Surfaces

Selected Plenary Lectures of the IUPAC-Conference on
Colloid and Surface Science in Budapest, September 15–20, 1975

Edited by

Prof. Dr. F. HORST MÜLLER — Marburg

Prof. Dr. ARMIN WEISS — München

Prof. Dr. E. WOLFRAM — Budapest

With 129 figures and 23 tables

Springer-Verlag Berlin Heidelberg GmbH **1976**

ISBN 978-3-662-16079-4   ISBN 978-3-7985-1801-8 (eBook)
DOI 10.1007/978-3-7985-1801-8

CONTENTS

Danielsson, I., J. B. Rosenholm, P. Stenius, and *S. Backlund* (Åbo/Finland), Lyotropic mesomorphism and aggregation in surfactant systems (with 13 figures) 1

Lipatov, Yu., S. (Kiev/USSR), Adsorption of macromolecules from concentrated solutions (with 17 figures and 5 tables) 12

Matijević, E. (New York/USA), Preparation and characterization of monodispersed metal hydrous oxide sols (with 20 figures) 24

Mirnik, M. and *S. Musić* (Zagreb/Yugoslavia), Adsorption of iodide ions and nucleation of freshly prepared silver iodide sols (with 7 figures) 36

van Olphen, H. (Washington/USA), Clay-water relationship — theory and application. A review (with 4 figures) 46

Hsing, H. H. and *A. C. Zettlemoyer* (Bethlehem/USA), Water on silica and silicate surfaces. IV. Silane treated silicas (with 18 figures and 5 tables) 54

Sørensen, T. S., M. Hennenberg, A. Steinchen-Sanfeld, and *A. Sanfeld* (Lyngby/Denmark and Brussels/Belgium), Surface chemical and hydrodynamic stability (with 19 figures) . . . 64

Shchukin, E. E. and *E. A. Amelina* (Moscow/USSR), Cohesion of particles in disperse systems (with 5 figures and 4 tables) 71

Shinoda, K. (Yokohama/Japan), Dissolution due to the orientation, arrangement and structure formation of molecules (with 9 figures and 1 table) 80

Sonntag, H. and *H. Pilgrim* (Berlin/DDR), Stern potential, zeta potential and dipole moment of aerosil particles dispersed in electrolyte solutions (with 3 figures) 87

Tamai, Y. (Katahira/Japan), Surface energy analysis of solids and its application (with 3 figures and 4 tables) 93

Baszkin, A., M. Deyme, M. Nishino, and *L. Ter-Minassian-Saraga* (Paris/France), Surface chemistry and wettability of modified polyethylene (with 12 figures and 4 tables) 97

PROGRESS IN COLLOID AND POLYMER SCIENCE

Fortschrittsberichte über Kolloide und Polymere

Supplements to "Colloid and Polymer Science" · Continuation of „Kolloid-Beihefte"

Vol. 61   1976

Progr. Colloid & Polymer Sci. **61**, 1—11 (1976)
© 1976 by Dr. Dietrich Steinkopff Verlag GmbH & Co. KG, Darmstadt
ISSN 0340-255 X

Plenary lecture of the IUPAC-Conference on Colloid and Surface Science in
Budapest, September 15—20, 1975

Department of Physical Chemistry, Åbo Akademi, Åbo, Turku (Finland)

Lyotropic mesomorphism and aggregation in surfactant systems

I. Danielsson, J. B. Rosenholm, P. Stenius, and *S. Backlund*

With 13 figures

(Received December 9, 1975)

Introduction

The nature of the interactions between amphiphilic compounds and simple solvents that result in very complex association equilibria found in such systems has been very extensively studied. The methods used include thermodynamic investigations of the aggregation numbers and the energetics of the association processes, determinations of aggregate structures from the flow properties of the solutions and direct studies of the interaction between the associated molecules and between them and the solvent by spectroscopic methods. From a classical chemical point of view surfactant systems are highly interesting in that they represent a case where, by variation of the concentration, a series of reversibly formed aggregates are obtained from small, simple complexes in solution to multimolecular, indefinitely large aggregates. Let us first of all give a qualitative survey of our picture of this aggregate formation (fig. 1).

The existence of small complexes (pre-micelles) formed by a few surfactant ions in aqueous solutions below the critical micelle concentration, c.m.c. (fig. 1, L_1) was postulated many years ago. Above the well-defined c.m.c., ordinary Hartley micelles predominate (fig. 1, L_1).

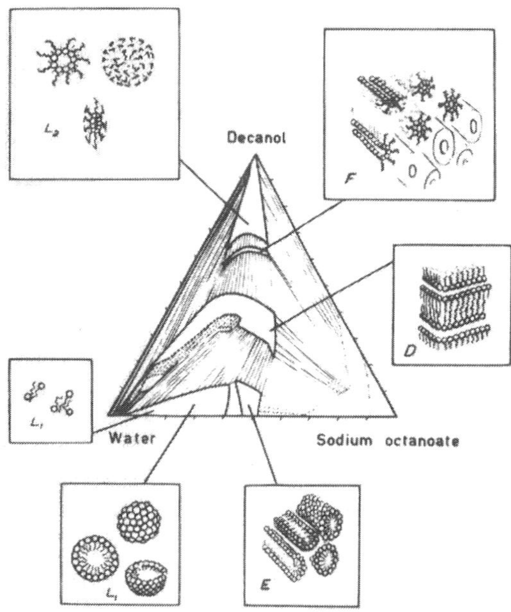

Fig. 1. Structures of different aggregates and phase equilibria at 20 °C in the system water-sodium-*n*-octanoate-*n*-decanol (1);

L_1 premicelles and Hartley micelles in isotropic water-rich solution L_1,

L_2 "inverted" micelles in isotropic oil-rich solution L_2,

D   lamellar mesoaggregates,

E   hexagonally ordered hydrophilic rods,

F   hexagonally ordered hydrophobic rods

If the concentration is increased above the solubility limit, various kinds of lyotropic liquid crystals (or lyotropic mesophases) are formed in which there is no longer an aqueous continuum. The same effect can be brought about by the addition of a weakly polar additive, even in quite dilute solutions.

The term "lyotropic mesomorphism" has often been used to describe the formation of thermodynamically stable liquid crystalline phases through the penetration of the solvent between the lattices of a (semi)crystalline amphiphile. These mesophases are built up of multimolecular units that often are called mesoaggregates, with a structure very similar to that of micelles. They may, therefore, have an inner, lipophilic moiety that is surrounded by an intervening aqueous polar layer completely bound in the phase (1). Examples of the structure of these mesoaggregates are shown in fig. 1, the lamellar phase D, and the phase E which is built up of hexagonally arranged hydrophilic rods. Under the influence of outer, orienting forces, these mesoaggregates may be of indefinite length in one or two dimensions. However, there are indications that some mesophases may contain different kinds of nearly isodimensional aggregates. Their structures, however, are still under discussion (2). A very common type is phase F (fig. 1), consisting of hexagonally arranged hydrophobic rods with a weakly polar interstitial liquid. The structure of these mesoaggregates corresponds to the "inverted micelles" (fig. 1, L_2) with polar, water-binding inner parts, that occur in weakly polar or completely non-polar solvents. It has been well established during recent years that the so-called microemulsions are actually equilibrium solutions of this type.

It is important to note that ordinary, as well as inverted micellar aggregates exist only in true solution where they are randomly distributed in the solvent, whereas the mesoaggregates form continuous liquid crystalline structures with ordered lattices, whose interplanar distances vary only to a limited degree which is determined by the composition and the temperature of the system.

Thus, the mesomorphic phases are characterized by a "long range order" that is lacking in micellar solutions; the "short range order" is fairly similar in micelles and mesoaggregates.

It was shown by *Ekwall* and his co-workers that mesophases in surfactant systems are true equilibrium phases that obey the phase rule (fig. 1). Their structure was clarified by *Luzzati* and his group (3). Our work has been aimed at a more quantitative description of the criteria for the formation of all these different types of aggregates, that is, we have tried to establish a clear-cut picture of the delicate balance between different attractive and repulsive forces that play a role in these systems. Such a picture is of the greatest interest: technologically for the formulation of emulsions and the possible use of lyotropic liquid crystals in applications now limited to thermotropics and biologically since it is obvious that this same type of balance plays a crucial role in the formation of biological membranes. In this lecture we will try to elucidate the question of the forces leading to the very first steps of association, premicelles; then we will discuss some of the thermodynamic properties of the higher aggregates and, finally, we will present some new spectroscopic data for the aggregated systems.

Pre-micellar association

To avoid the difficulties posed by the low concentrations below the c.m.c. for long-chain compounds we have investigated the first steps of association for short-chain carboxylates in solutions to which an inert electrolyte (sodium chloride) was added to keep the total ionic strength of the solutions constant. If the concentration dependence of the activity coefficients can be mastered in this way (we will return to this point later), the complex formation can be quantitatively investigated by studying the acid-base equilibria of the carboxylate complexes using the methods developed by *Sillén* (4). Potentiometric titrations of the carboxylates at different total concentrations are described in terms of the quantity Z (fig. 2), the total number of protons bound per carboxylate ion. This quantity is easily shown to be a unique function of pH or the negative logarithm of the hydroxyl ion concentration, pOH, and the stability constants of the aggregates that are formed in the solution; the shape and location of the curves of Z against pOH tell us which complexes are present and it is also easy to assert how various experimental errors would affect our conclusions (5, 6). In fig. 2, above

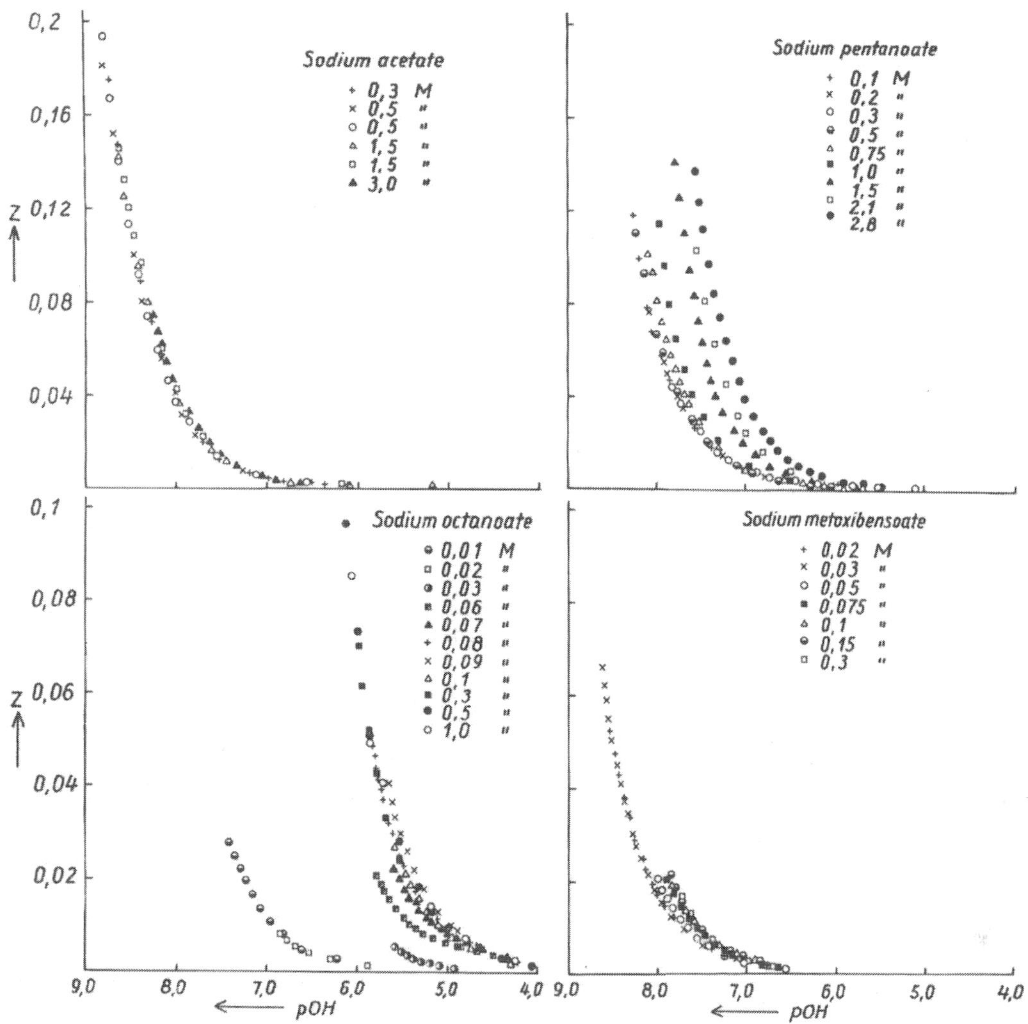

Fig. 2. The hydrolysis of carboxylate ions described as the number (Z) of protons bound to an anion in the "mean complex" at different molar concentrations. Temperature 25 °C, concentration of sodium ions (NaCl) 3 M (5, 13)

to the left, are shown the Z curves for different concentrations of sodium acetate: they are quite independent of concentration which indicates that a single simple equilibrium, the formation of acetic acid, predominates. For higher homologues, from the butyrate upwards, Z is displaced towards lower pOH values as the concentration increases. Above to the right in fig. 2 are shown the Z curves for sodium pentanoate. These curves can be explained on the basis of the assumption that the complexes HB, H_2B_4, H_3B_4 and HB_{11} are formed; their stability constants we have adjusted to give the „best possible" agreement with the experimental data (on a least squares deviation basis). The same methods have been applied to higher homologues

up to sodium decanoate, where difficulties are encountered due to the very low solubility of decanoic acid below the c.m.c. The results for sodium octanoate are shown in fig. 2, below left. It can be seen that at high concentrations the Z curves are almost independent of the octanoate concentration, the shape of the curves being that for a simple mononuclear equilibrium. The stability constant, however, is quite different from that found for octanoic acid at low concentrations. The Z curves can be explained assuming the complexes HB, HB_5, HB_9, B_9; the dissociation constants of the carboxyl groups in the large aggregate are all equal, which explains the coinciding Z curves. Micellar aggregates predominate above a fairly well-

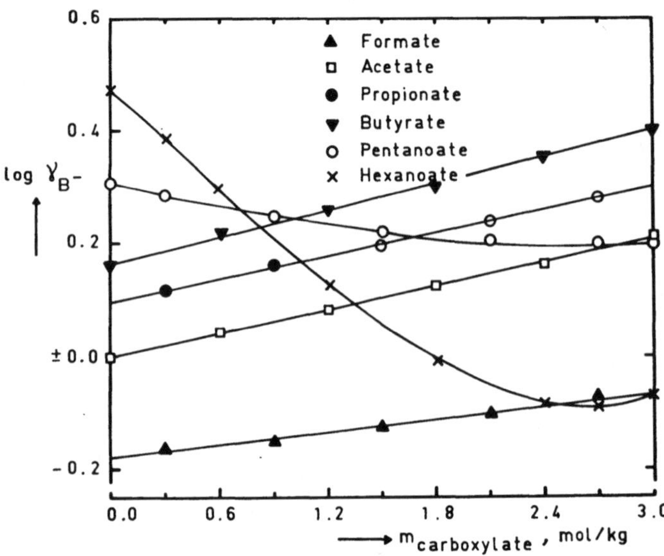

Fig. 3. Activity coefficients of carboxylate ions at different molar concentrations calculated from activities of sodium chloride ions and water. Temperature 25 °C, sodium ion concentration 3 molal (NaCl) (9);

a) formate   d) butyrate
b) acetate   e) pentanoate
c) propionate   f) hexanoate

defined concentration limit, 0.05 M. The method cannot significantly distinguish between complexes with aggregation numbers ranging from 9 to 17. Higher aggregation numbers than that are, however, highly improbable (7).

The same method has been used to study the influence of a weakly polar solubilizate, n-decanol, on the association in sodium pentanoate solutions (8). The aggregation number of the anions remains constant on solubilization; the effect of solubilization is an increased stability of the micelles resulting in a shift of the Z curves.

It is, of course, possible that the high ionic strength causes the formation of aggregates and bonds of a type quite different from those in additive-free solutions. Also, the assumption that the activity coefficients remain constant is crucial to the calculations. We have studied these coefficients by potentiometry, gas chromatography and vapour pressure osmometry. In fig. 3 is shown how V_B-changes with the carboxylate molality, in the ionic mechanism 3 m NaCa. For different mononuclear species these changes, however, compensate each other in the equilibria (10). This is also clearly seen from the Z curves of non-associating compounds, e.g., the acetate, which do not change their position as the concentration of acetate increases. The higher homologues all show a region of non-shifting Z-curves at low concentrations. We are thus confident that the shift in the Z curves really does indicate an association. However, since the activity of the free carboxylate ions is raised into powers higher than one in the stability constants of the polynuclear complexes, the actual aggregation numbers remain somewhat uncertain.

A combination of gas chromatographic determinations of the fatty acid activity and determinations of the pH of the solutions makes it possible to investigate the activity coefficients of the carboxylic ions in systems where they associate. Fig. 4 shows pentanoate ion activities in 3 M NaCl (11). The activity increases linearly up to the region where the Z curves (fig. 2) indicate association, which confirms our conclusions.

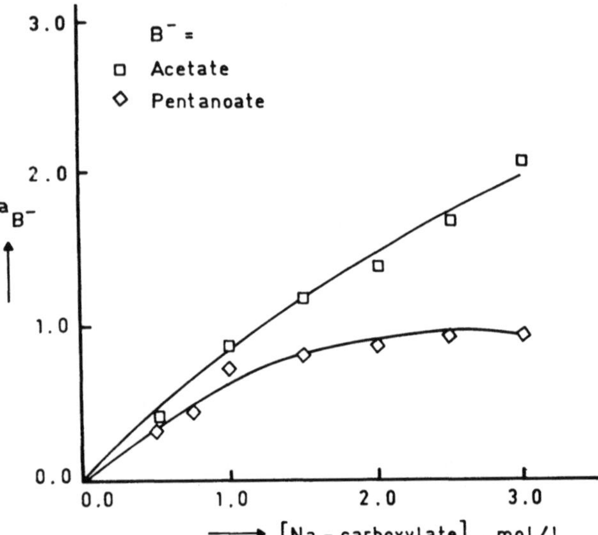

Fig. 4. Activities of acetate and pentanoate ions calculated from vapour pressure of fatty acid and hydrolytic data. Temperature 25 °C, molar concentrations and 3 M concentration of sodium ions (NaCl) (11)

The occurrence of small aggregates may also be qualitatively confirmed by measurement of the densities and vapour pressures of these solutions of short-chain carboxylates, as well as by calorimetric determination of the heats of mixing of 3 m NaCl and 3 m sodium carboxylate. Fig. 5 shows the apparent molar enthalpy of the carboxylates. It is larger if they associate than if they were molecularly dispersed. The association apparently is "entropy governed", the binding in pre-micelles as well as in micelles being of a hydrophobic nature.

The non-polar characteristics of the binding in pre-micellar aggregates is also evident from comparison with the complex formation of aromatic hydroxy- and methoxy carboxylates caused by polar forces. Thus, there is association in sodium salicylate solutions in 3 M NaCl at concentrations considerably lower than those at which straight-chain carboxylates with a corresponding number of CH_2 groups associate (12). This is even more evident for sodium 2-methoxy bensoate (fig. 2) (13). Only the complexes HB and HB_2 have to be assumed to describe these curves; the double ions have a very high stability constant and are evidently held together by polar forces. No higher complexes, as formed by, for example, the aliphatic sodium pentanoate, are found.

Ionic amphiphiles with aromatic rings of the type mentioned above are able to solubilize additives that are not soluble in pure water. Since these so-called "hydrotropes" obviously do not form micelles, the solubilization mechanism must be quite different from that of association colloids proper.

Properties of surfactant systems containing higher aggregates

a) Chemical potentials in the ternary system water-sodium n-octanoate-n-pentanol

It is possible to investigate normal Hartley micelles, mesoaggregates and inverted micelles at a defined temperature in one single ternary system of water, surfactant and a third, weakly polar component. Thus, *Ekwall* has studied the water activities in all the phases formed at 25 °C in the system water-sodium octanoate-decanol (14). A thermodynamic description of the system requires knowledge of the activity of at least one of the other components, too. We have developed a method to determine the

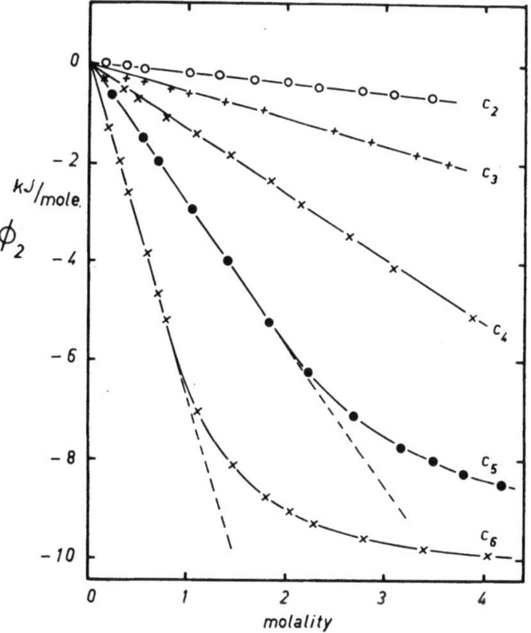

Fig. 5. The relative apparent molar enthalpies of sodium alkanoates in the system 3 m RCOONa — 3 M NaCl. Temperature 25 °C, standard state of RCOONa infinitely diluted alkanoate in 3 M NaCl (5)

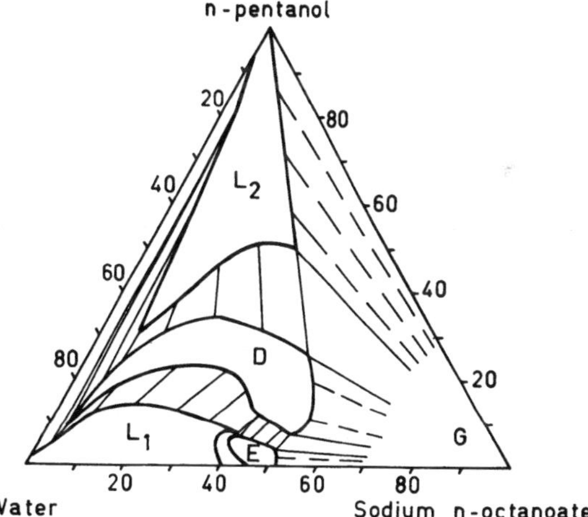

Fig. 6. Phase equilibria in the system water-sodium-*n*-octanoate-*n*-pentanol at 20 °C (15)

activity of the weakly polar component by gas chromatography. To simplify these measurements, we have chosen a third component with a higher vapour pressure, i.e., the system water-sodium octanoate-*n*-pentanol. The phase diagram of this system has been determined by *Ekwall* and *Mandell* at 20 °C (fig. 6) (15). In this system, the only mesophases formed are the structurally well-known lamellar phase D and the hexagonal phase E. In addition, an aqueous solution L_1 and a pentanolic solution L_2 occurs.

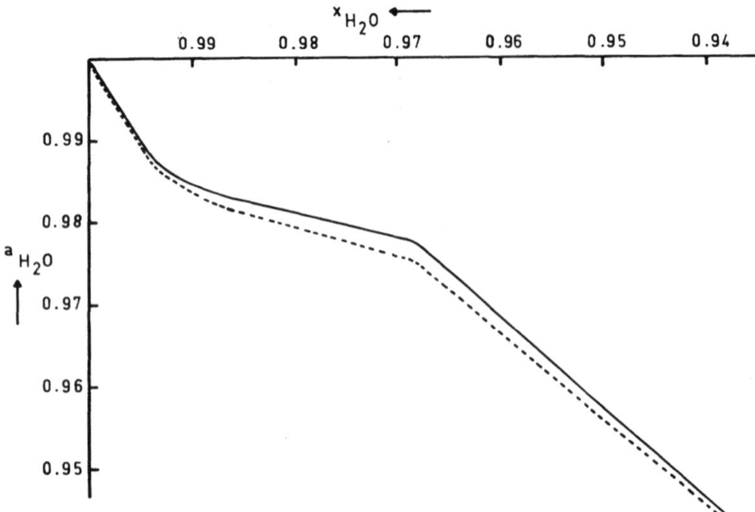

Fig. 7. The activity of water in octanoate solutions at 25 °C as a function of its mole fraction. The unbroken line corresponds to the two-component system H_2O-NaC_8, the broken line to octanoate solutions saturated with pentanol (18)

The activity of water in aqueous sodium octanoate solutions is only slightly decreased at the c.m.c. compared with nonassociating electrolytes (fig. 7) (16, 17, 18). At very high octanoate concentrations the water activity again starts to decrease more rapidly, at a concentration somewhat higher than that sometimes defined as "the second critical concentration".

As shown by the dotted line in fig. 7, the vapour pressure is slightly lowered by solubilized pentanol.

The activity of water in aqueous solutions of octanoate and pentanol that are in equilibrium with lamellar phase D or hexagonal phase E ranges from .94 to .99 at 25 °C. These high values agree with those found by *Ekwall* with decanol as the solubilizate. The vapour pressure of water is relatively high everywhere in the D and E phases. It has been pointed out by *Ekwall* that the small changes noted can be understood in terms of the ability of the polar groups in the

mesoaggregates to bind water: part of the water is intercalated between the mesoaggregates and part of it is strongly bound to polar groups and ions. The decrease in activity with increasing content of surfactant or solubilizate reflects the decreasing amount of intercalated water; the smallest water content giving a stable mesophase corresponds roughly to that required to hydrate the polar groups and ions.

Fig. 8 shows the activity of pentanol in sodium octanoate solutions as a function of the mole fraction of pentanol at various constant octanoate concentrations (18). The standard state is pure pentanol. With very small additions of octanoate, the activity of pentanol (within the accuracy of our experiments) follows Henry's law, that is, the octanoate has no effect on its activity. However, deviations towards lower pentanol activities are found at octanoate concentrations well below the c.m.c., indicating the formation of mixed complexes of pentanol

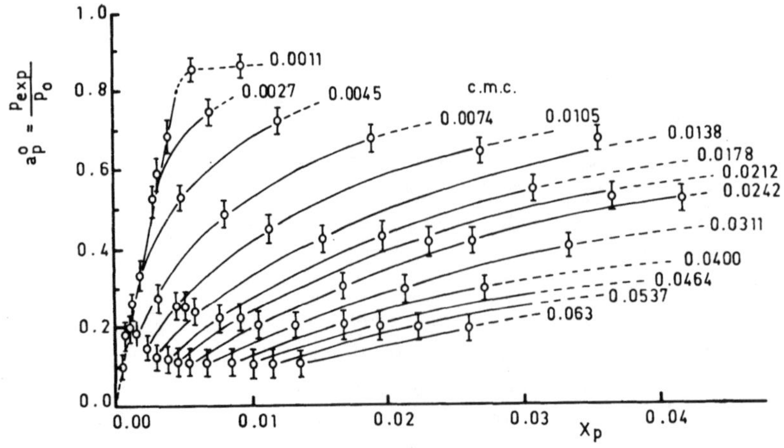

Fig. 8. The activity of pentanol at 25 °C in isotropic solutions L_1 of octanoate in water solubilizing pentanol. The abscissae are mole fractions of pentanol and the curves correspond to different ratios of moles of sodium octanoate per mole of water (18)

and octanoate (i.e., an effect similar to that found in the potentiometric titrations of sodium pentanoate described above). This effect is similar for increasing octanoate concentrations in most of the concentration regions above the c.m.c. At very low concentrations, the pentanol roughly follows Henry's law, but as the amounts are increased, strong deviations are found. For high concentrations of pentanol the activities again become linearly dependent on the mole fraction of pentanol, especially at high octanoate contents. The slope of these lines, however, is much less than predicted by Henry's law. Thus, when pentanol is added to a micellar sodium octanoate solution, there is little interaction between octanoate and pentanol at very low additions, while the solubilization of pentanol at higher concentrations can be described in terms of a distribution between the micelles or a micellar "pseudophase" and the surrounding solution. The distribution coefficient for pentanol between micelles and solution is rather high, about 18. From the distribution coefficient it is also possible to estimate the volume fraction of the micellar pseudophase: for example, in a 2 M sodium octanoate solution it is about 0.7. This implies that about 10 water molecules per molecule of octanoate are bound to the "pseudophase" that solubilizes pentanol; this value is of the same magnitude as the hydration number 9.5 found in diffusion measurements by *B. Lindman* (19).

b) Calorimetric investigations

The activity measurements described above have been supplemented by calorimetric investigations of excess enthalpies. The excess enthalpy of pure sodium octanoate in micellar solution is shown in fig. 9. The standard states are pure water and sodium octanoate at infinite dilution. The c.m.c. is indicated by a slight change in the slope of the curves, but as the formation of micelles begins to predominate, the enthalpy curve deflects sharply downwards. The enthalpy of transfer of sodium octanoate to the hexagonal phase E is only about 0.3 kJ/mol. As the water content of the phase is decreased, the enthalpy of the octanoate remains constant at first, but close to the highest concentrations of octanoate there is a sharp decrease. This probably reflects changes in the binding of water to the ionic end-groups or, possibly, a depletion

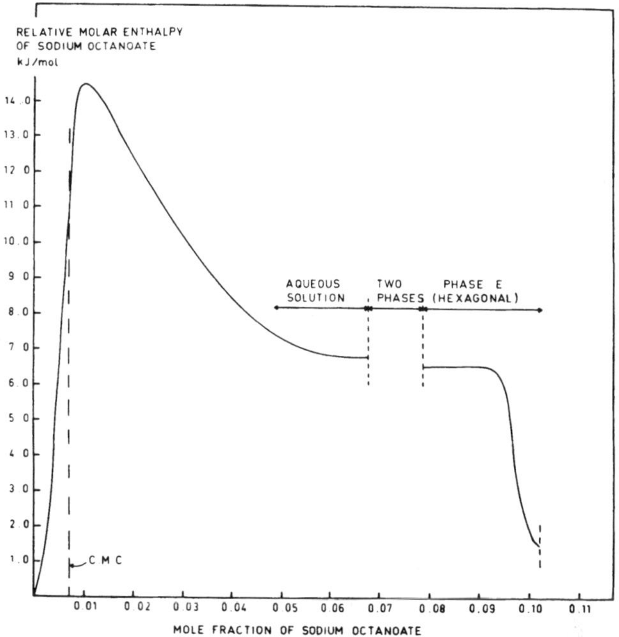

Fig. 9. The partial excess enthalpy of sodium octanoate in the system water-sodium octanoate as a function of the molar fraction of the salt at 25 °C (18)

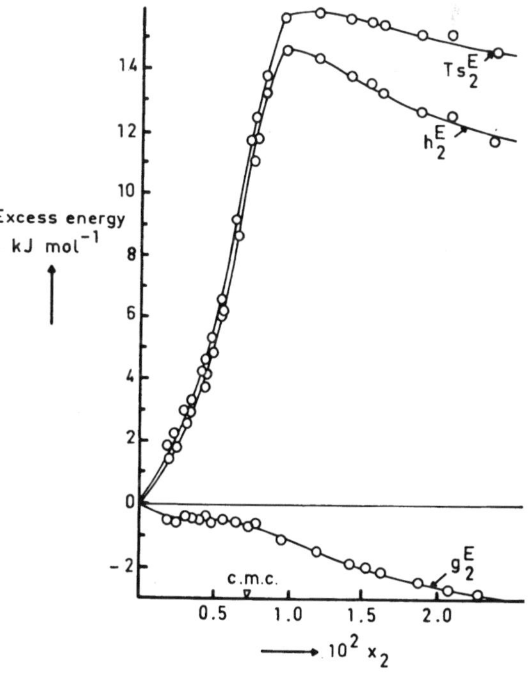

Fig. 10. Partial excess quantities of sodium octanoate in water solutions. The abscissae are molar fractions of solute at 25 °C (18)

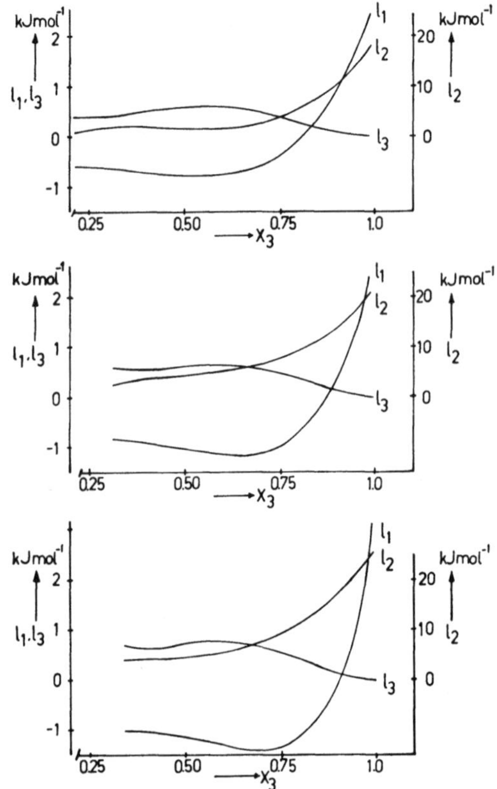

Fig. 11. Partial excess enthalpies of water (l_1), sodium octanoate (l_2) and n-decanol (l_3) in isotropic oil phase L_2 in one three-component system. From top to bottom the curves correspond to the ratios 26.28, 10.83 and 6.68 moles of water per mole of octanoate (20)

of the "intercalating" water. Thus, substantial structural changes are found in homogeneous solutions as well as in mesophases.

Fig. 10 gives the same excess enthalpies, as well as the excess free energy g_2^E of the octanoate, as calculated from water activities. The excess energies are weakly negative and their absolute values are much smaller than the enthalpies. The excess entropy is given as $Ts_2^E = h_2^E - g_2^E$ in the uppermost curve in fig. 10. The interpretation of curves of this type has led to the introduction of concepts like "hydrophobic hydration" and "hydrophobic bonding" as the driving force behind micelle formation. The unique interaction between hydrocarbon chains and water is certainly important, as seen, for example in the strong entropy effects below the c.m.c. However, we feel that great care should be taken in any

formulation of a more detailed molecular model of micelle formation: many more curves of this type for several homologues, beginning with the lowest, non-associating compounds, and for several different end-groups should be known before it is possible to give a model with truly predictive not just descriptive properties. We are at present conducting many experiments to collect more data of this type.

Similar methods can be used to study the association processes in all the different phase regions in ternary systems. Fig. 11 shows an example of determinations of partial molar enthalpies in solutions of inverted micelles: we have studied the L_2 region of the system water-sodium octanoate-decanol at certain constant ratios water/octanoate. The molar enthalpy of the octanoate is $0-5$ kJ mol^{-1} throughout the region where there is enough water for primary hydration of the ions in the inner polar part of the inverted micelles. In this relatively water-rich region of L_2 the partial molar enthalpies of the water and the decanol, respectively, are relatively small and compensate each other almost completely. In the inverted micelles the first molecules of water are extremely strongly bound.

c) Spectrometry

In the formulation of molecular pictures of the structures in surfactant systems it is extremely important that the thermodynamic results are supplemented by spectroscopic data.

Using low-angle X-ray scattering techniques we have estimated that the radius of octanoate micelles in water is about 2.5 nm (21, 22). The light scattering technique developed by *Vrij* and *Huisman* (23) yields micellar weights of about 4200 g/mol, which is in fair agreement with the X-ray results. This micellar weight is $50-60\%$ higher than that calculated directly from the aggregation numbers found potentiometrically. However, if the counter ion binding and the hydration determined by other methods are taken into account, the micellar radii calculated from different determinations agree to within 15% (7).

The nature of the binding in the aggregates can, in principle, be studied by infra-red and Raman spectroscopy as well as by nuclear magnetic methods. Due to the dominating presence of water in most surfactant systems, i.r. techniques can be used only in the phase with the lowest water content, L_2, i.e., solutions of inverted micelles in lipophilic solvents. Fig. 12 shows the H_2O-OH stretching band (3500 cm^{-1}) as a function of the reciprocal molality of water in the same solutions as those studied in the enthalpy measurements above (24). There is a close correlation between the regions that show anomalous effects here and in the enthalpy measurements. As the molar ratio water/octanoate is increased in the region of L_2 with the highest water content, the H_2O-OH stretching frequency is displaced towards lower energies. The shift towards higher energies is on the other hand strongest for the lowest molar ratios water/octanoate.

Thus, i.r.-spectrometric as well as calorimetric results indicate that association phenomena analogous to micelle formation in the aqueous L_1 phases take place in the L_2 phases, too. The changes in the L_2 phases are less sudden, indicating either a lower aggregation number or a more stepwise association.

The structural state of the hydracarbon chain can be investigated by Raman spectroscopy according to *Lippert* and *Peticolas* by measuring the ratio between the intensities of "gauche" band at about 1090 cm^{-1} and the "trans" bands at about 1065 cm^{-1} and 1125 cm^{-1} (25). Fig. 13 shows that the state of the hydrocarbon chain in sodium octanoate is shifted strongly towards the trans state, that is, towards a crystalline state, as the c.m.c. is reached in aqueous solution. Similar conclusions can be reached on the basis of nmr investigations by *B. Lindman* and *T. Drakenberg* (26). An extensive crystalline structure in the micelles indicates that the total entropy increase on micelle formation is caused by the destruction of water cages around the molecularly disperse hydrocarbon chains, and not by an increased torsion in the micelles (27, 28).

Larsson and *Rand* (39) have pointed out that a comparison of the band at 2940 cm^{-1} (generally accepted to be due to asymmetric vibration of CH_2 groups) with the other bands in the CH-stretching vibration region gives information on the environment character of the hydrocarbon chains, that is, their contact with water. Fig. 13 shows an investigation of the changes in the polar contacts of hydrocarbon chains with

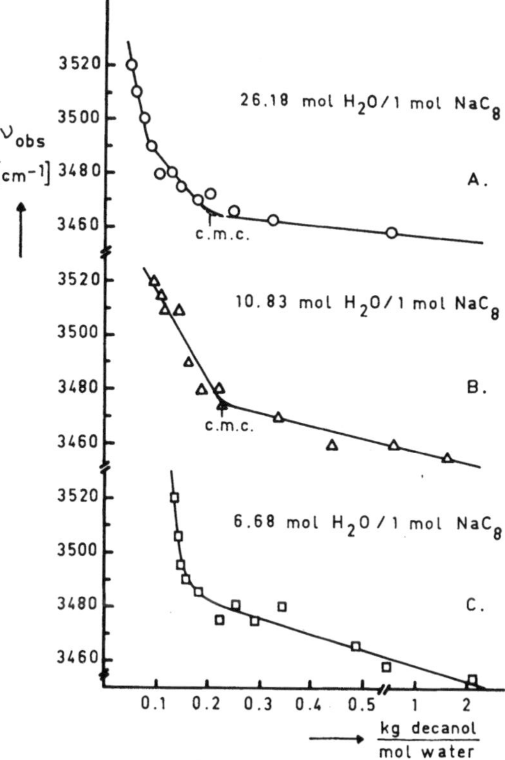

Fig. 12. The wave number of the water OH-stretching band in the system water-octanoate-pentanol as a function of the inversed concentration of decanol expressed as kg decanol per moles water. The curves refer to different ratios of moles of water per mole of octanoate (34)

concentration in octanoate solutions. The c.m.c. does not show up very clearly, but it is obvious that the hydrocarbon chains are transferred from a dominating aqueous environment below the c.m.c. to a more lipid-like state above the c.m.c. These are just a few examples how the Raman spectroscopy has been used to investigate association colloids.

Conclusions

Pre-micelles formed by short-chain carboxylates as well as micelles proper are formed by hydrophobic interactions and represent different steps in a continuous association process leading to an increased crystallinity of the hydrocarbon chains. Since the heat effects on formation of lamellar or rod-like mesoaggregates are small, entropy effects are of dominating

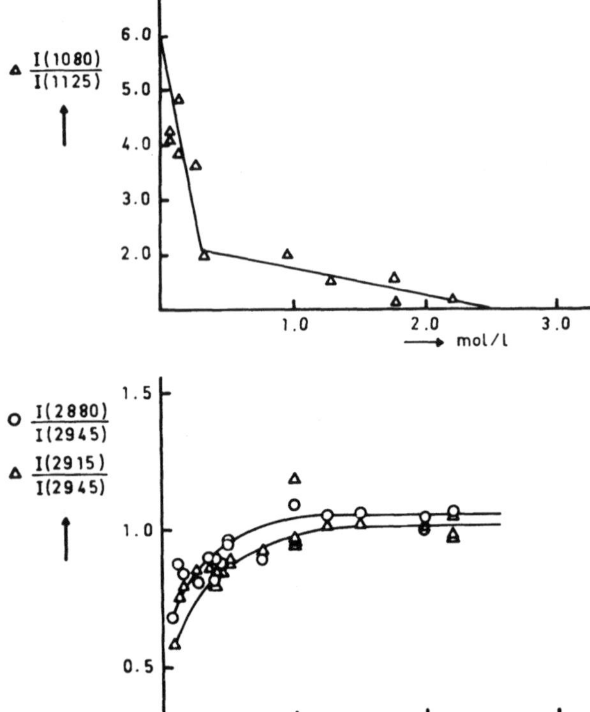

Fig. 13. Intensity ratios of the C-C stretching vibration bands at 1080 cm^{-1} and 1125 cm^{-1}, the C-H stretching vibration bands at 2880 cm^{-1} and 2945 cm^{-1} and at 2915 cm^{-1} and 2945 cm^{-1} respectively of the Laser-Raman spectra of aqueous octanoate solutions at different molar concentrations

importance here, too. A successive restructuring from lower to higher aggregates as the concentration of amphiphile increases seems natural, but the equilibrium measurements, of course, do not give any direct knowledge about reaction mechanisms.

In the molecularly disperse concentration region of the "oil phase" L$_2$ the hydrocarbon chain of the amphiphile as well as the ionic groups are surrounded by the lipophilic solvent. The partial molar enthalpies of the water and the amphiphile indicate that the hydration of the ionogenic groups in the inner part of the inverted micelles constitutes the main part of the energy of association. The same factors are obviously also decisive for the formation of the rod-like mesoaggregates in phase F, which, in the phase diagrams, is always found to be in equilibrium with inverted micelles or lamellar mesoaggregates only. The latter are formed when

the polar and non-polar forces are roughly equal. It would seem that the experimentally found thermodynamic effects can be divided into lipophilic and hydrophilic contributions which make it possible to create quantitative criteria for the formation of various mesophases in a way somewhat similar to the R-theory suggested by *Winsor* (2).

Our results give some quantitative support to a very much simplified classification of the association processes in the following way: In aqueous solutions, hydrophobic interactions lead to the transfer of completely hydrated ionic groups to pre-micelles, Hartley micelles or mesoaggregates with a lipophilic inner moiety. In lipophilic solutions, the interaction between the ionic groups and the water lead to the formation of inverted micelles or mesoaggregates with a hydrophilic inner part.

References

1) *Ekwall, P., I. Danielsson*, and *P. Stenius*, Surface Chemistry and Colloids, MTP Intern. Rev. Sci. Phys. Chem. Ser. 1, Vol. 7, p. 97—145 (London 1972); *Ekwall, P.* and *P. Stenius*, Phys. Chem. Ser. **2**, Vol. 7, p. 215—248 (London 1975).
2) *Gray, G. W.* and *P. A. Winsor*, Mol. Cryst. Liquid Cryst. **26**, 305 (1974).
3) *Luzzati, V.* and *A. Tardieu*, Ann. Rev. Phys. Chem. **25**, 79 (1974).
4) *Ingri, N., E. Lagerström, M. Frydman*, and *L. G. Sillén*, Acta Chem. Scand. **10**, 1034 (1957).
5) *Danielsson, I.* and *P. Stenius*, J. Colloid Interface Sci. **37**, 264 (1971).
6) *Stenius, P.*, Acta Chem. Scand. **27**, 3897 (1973).
7) *Friman, R., K. Pettersson*, and *P. Stenius*, J. Colloid Interface Sci. **53**, 90 (1975).
8) *Stenius, P.* and *L. A. Filén*, Progr. Colloid & Polymer Sci. **56**, 21 (1975).
9) *Backlund, S.* and *K. Palm*, Acta Chem. Scand. A **28**, 595 (1974).
10) *Danielsson, I.* and *P. Stenius*, Trans. Royal Inst. Techn., Stockholm, Sweden **254**, 81 (1972).
11) *Backlund, S.* and *I. Danielsson*, Chemie, physikalische Chemie und Anwendungstechnik der grenzflächenaktiven Stoffe, Vol. II, 1013—1021 (München 1973).
12) *Stenius, P.*, Unpublished data.
13) *Halling, K.*, Unpublished data.
14) *Ekwall, P.*, Liquid Crystals and Ordered Fluids, Vol. 2, p. 177 (New York 1974).
15) *Ekwall, P., L. Mandell*, and *K. Fontell*, Mol. Cryst. Liquid Cryst. **8**, 157 (1969).
16) *Robinson, R. A.* and *R. H. Stokes*, Electrolyte Solutions, 2nd ed., 484 (London 1959).
17) *Ekwall, P., H. Eikrem*, and *P. Stenius*, Acta Chem. Scand **21**, 1639 (1967).

18) *Rosenholm, J. B.* and *P. Stenius*, To be published.
19) *Lindman, B.* and *B. Brunn*, J. Colloid Interface Sci. **42**, 388 (1973).
20) *Bedö, S.*, To be published.
21) *Reiss-Husson, F.* and *V. Luzzati*, J. Colloid Interface Sci. **21**, 534 (1966).
22) *Svens, B.* and *J. B. Rosenholm*, J. Colloid Interface Sci. **44**, 495 (1973).
23) *Vrij, A.* and *J. Th. G. Overbeek*, J. Colloid Interface Sci. **17**, 570 (1962); *Huismann, H. F.*, Proc. Kem. Ned. Akad. Wetensch. Ser. B **67**, 367 (1964).
24) *Rosenholm, J. B.*, To be published.
25) *Lippert, J. L.* and *W. L. Peticolas*, Biochim. Biophys. Acta **282**, 8 (1972).
26) *Drakenberg, T.* and *B. Lindman*, J. Colloid Interface Sci. **44**, 184 (1973). Personal communication.
27) *Nemethy, G.* and *H. A. Sheraga*, J. Phys. Chem. **66**, 1773 (1962).
28) *Aranow, R. H.* and *L. Witten*, J. Phys. Chem. **46**, 1643 (1960).
29) *Larsson, K.* and *R. P. Rand*, Biochim. Biophys. Acta **326**, 245 (1973).

Authors' address:

Ingvar Danielsson, Jarl B. Rosenholm, P. Stenius,
and *S. Backlund*
Department of Physical Chemistry, Åbo Akademi,
Porthansgatan 3—5
SF-20500 Åbo, Turku 50, Finland

Progr. Colloid & Polymer Sci. **61**, 12–23 (1976)

ISSN 0340-255 X

Plenary lecture of the IUPAC-Conference on Colloid and Surface Science in
Budapest, September 15–20, 1975

The Institute of Macromolecular Chemistry at the Academy of Sciences of the Ukrainian SSR

Adsorption of macromolecules from concentrated solutions

Yu. S. Lipatov

With 17 figures and 5 tables

(Received December 9, 1975)

The processes of polymer adsorption from solutions on solid surfaces are of great interest from the point of view of developing concepts about the adhesion, structure and properties of polymeric composite materials. The most important properties of the latter are determined by the interphase adsorption interactions at the interface polymeric solution—solid body. The majority of works in the field of polymer adsorption are devoted to the study of diluted solutions. The development of the theory of adsorption from diluted solutions is based on the existing solution theories. This permitted to develop theoretic ideas about the specific features of adsorption of macromolecules linked with conformation of chains in solutions and about the peculiarities of the structure of the adsorption layers. These points have been considered in detail in our monograph (1).

The adsorption of polymers from concentrated solutions has been studied considerably lesser. This can be explained both by experimental difficulties arising in such investigations and the absence of the quantitative theory of concentrated solutions impairing the interpretation of the obtained results. Meanwhile, it is the adsorption from concentrated solutions that is of the greatest theoretical and practical interest since it permits to approach the study of structure of adsorption layers in the condensed phase.

At present sufficient experimental material has accumulated both with regard to the structure of concentrated polymer solutions and adsorption from them. So, we set forth a task to analyze the mechanism of polymer adsorption from concentrated solutions on the basis of existing concepts about their structure. We like to show that in this case the adsorption is predetermined by solution structure which also is a decisive factor determining the structure of adsorption layers.

As concentrated we shall consider solutions in which the statistic coils of macromolecules either interact with one another, forming intermolecular bonds, or begin to overlap which results in contraction of the coils and reduction of their size. As a criterium of concentrated solution in this instant a concentration can be used equal to approximately the reverse value of the characteristic viscosity.

The manifestation of forces of intermolecular interaction in solutions leads to formation in them of molecular associates or aggregates and other types of supermolecular structures in which molecule conformation can differ from the conformation in a diluted solution. The investigations of concentrated solutions brought about the conclusion that with the rise of solution concentration due to re-arrangement of intermolecular interaction forces, a transition is possible from the coiled conformations typical for diluted solutions in poor dissolvents to more uncoiled conformations. The reduction of coil size due to concentration rise and their overlapping can result in the transition of chains into a more uncoiled conformation which, on account of increased intermolecular interactions can thermodynamically become more advantageous. As the concentration of polymer solution rises, the processes of cross-linking run more and more intensely along with the onset of fluctuating thiotropic or stable structural net, as a result of this gel-formation in solutions is possible. The structure of polymer gel, as was anticipated in these papers and was subsequently conformed experimentally by the method of electronic microscopy, is formed as a result of continuous net onset, this net permeating the entire volume which is relatively unlikely for coiled macromolecules.

So, the changes of chain conformations due to deteriorated thermodynamic properties of the solvent will be dissimilar in diluted and concentrated solutions. The increase of intermolecular interactions in a concentrated solution and the appearance of molecular aggregates leads to other changes of chain conformations due to the change of surrounding atmosphere than in diluted solutions. Therefore, taking into account the contribution which the intermolecular interactions add to the properties of polymeric solutions, it has to be considered that the ideas about the behaviour of macromolecules in diluted solutions cannot be directly transferred to the concentrated solutions in which intense macromolecular interactions are realized. This should be borne in mind also when studying the data on adsorption of polymers from concentrated solutions by solid surfaces.

At present the polymeric solutions, even in cases when they are thermodynamically stable and are in the state of equilibrium, are not considered any more as systems in which molecular dispersity level is accomplished, but the existence of definite types of supermolecular structures is recognized. Their appearance is possible even in diluted solutions. Multiple experimental data show that the size of molecular aggregates can vary within a wide range, depending on the molecular weight, solution concentration, temperature, solvent quality, chain flexibility and other parameters. On account of this, the study of adsorption from solutions in which noticeable intermolecular interaction is observed calls for establishment of interrelation between the structure formation in solutions and the adsorption interaction with the adsorbent.

A series of our papers on the study of adsorption from moderately concentrated solutions for the first time put forward the idea about the transition of molecular aggregates onto the adsorbent surface and about the existence of a definite relationship between adsorption and structure formation. After that this concept was developed in the works of other authors who studied adsorption both of polymers and oligomers.

To obtain direct experimental proof of the macromolecular aggregates transferring from polymer and oligomer solutions onto the adsorbent surface, we resorted to the method of turbidity spectrum permitting determination of the number of aggregates in a unit of solution volume and their size. Let us examine the results.

Table 1. Dependence of aggregate sizes in solutions of oligoethylene glycol adipinate upon concentration of solution and quality of solvent at 20 °C

In acetone		In toluene	
Concentration C, g/100 ml	Size $\bar{r}_w$, Å	Concentration C, g/100 ml	Size $\bar{r}_w$, Å
0.500	100	0.500	400
1.000	300	1.000	700
1.500	400	1.500	1000
2.000	600	2.000	4000
3.000	800	3.000	6700
4.000	900		
5.000	1100		
8.000	1200		
10.000	1300		

Indicated in table 1 is the dependence of aggregate sizes of oligoethylene glycol adipinate molecules upon solution concentration and thermodynamic properties of the solvent.

Evidently, the solutions contain aggregates whose quantity and size depend upon solution concentration and solvent quality. As can be seen from table 1, the size of aggregates increases with the increase of solution concentration, — first slowly, then more intensely at concentrations exceeding 2.0 g/100 ml. The quality of solvent also noticeably effects the size of the aggregates. In a good solvent, — acetone, — the size of aggregates is 0.03 to 0.13 mμ, in a low-quality solvent, such as toluene, the size of aggregates increases to 0.08—0.70 mμ.

The structure formation in solutions of epoxy resins ЭД-5 and ЭДН-I was also studied. The dependence of size of epoxy resin aggregates represented in figs. 1 and 2 shows that the size increases regularly with the increase of solution concentration. It has to be noted that in resin

ЭДН-I, molecular weight 30000, the size of aggregates is by an order larger even at much lower concentrations than in resin ЭД-5, molecular weight 500. The size of the aggregates depends also upon the nature of the solvent. Fig. 1 shows that in low-quality solvent, — toluene, — the size of aggregates grows with the increase of solution concentration from 0.06

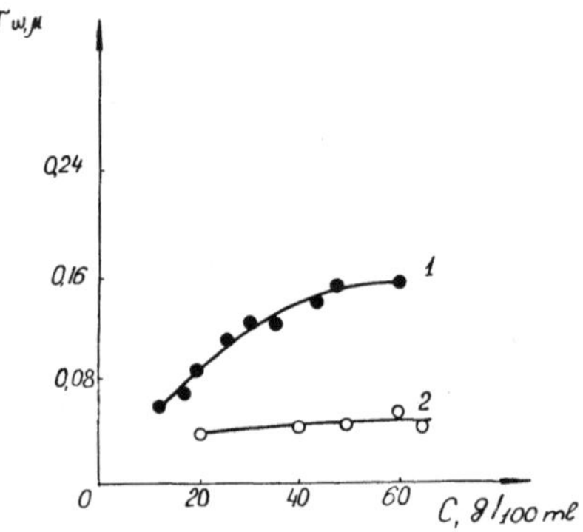

Fig. 1. Dependence of aggregate sizes of epoxy resin ЭД-5 upon solution concentration in toluene (curve 1) and dimethyl formamide (curve 2)

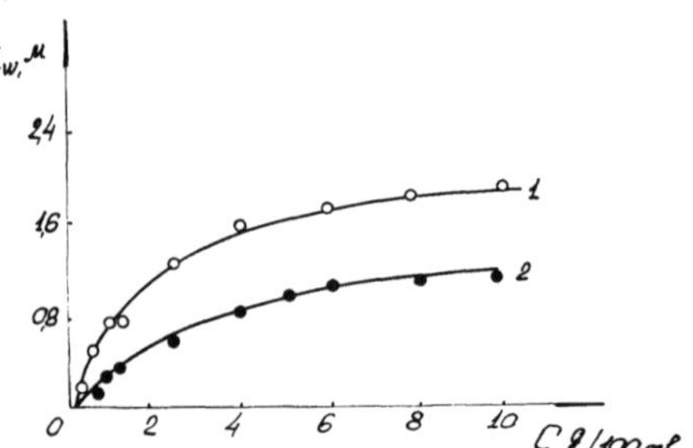

Fig. 2. Dependence of aggregate sizes of epoxy resin ЭДН-I upon solution concentration before (1) and after (2) adsorption (solvent: dimethyl formamide)

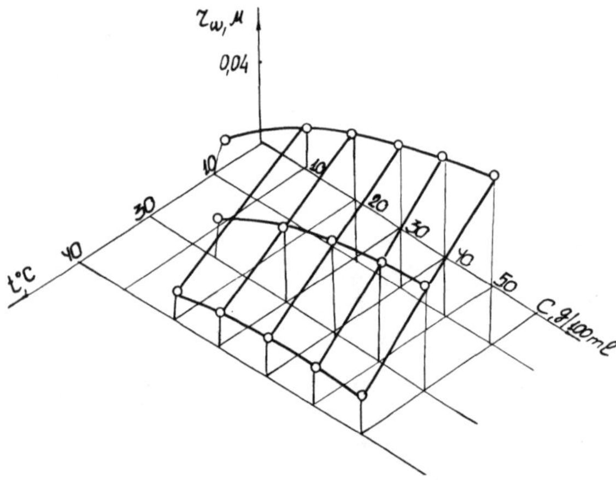

Fig. 3. Dependence of aggregate sizes of epoxy resin ЭД-5 upon solution concentration and temperature (solvent-toluene)

to 0.16 mμ, whereas in dimethyl formamide, — good-quality solvent, — the size of aggregates within the same range of concentrations does not change within the experimental error and comprises 0.04 mμ.

The effect of temperature on the size of aggregates is shown in fig. 3. Evidently, the rise of temperature to 40 °C leads to reduction of the size of aggregates and to the increase of their number.

So, in solution of ЭД-5 at $C = 15.0$ g/100 ml in toluene at 10° prevail aggregates measuring 0.14 mμ, their quantity being $0.80 \cdot 10^9$. The rise of temperature to 25° brings about reduction of the size of aggregates to 0.06 mμ with simultaneous increase of their number up to $1.0 \cdot 10^{10}$. At 40° the size of aggregates comprises 0.04 mμ and their number is $0.20 \cdot 10^{11}$.

Fig. 4 illustrates the effect of temperature on the size of aggregates for another polymer, — polycarbonate, — in the solution of dichlorethane. Here, as in the solutions of epoxy resin, the size of aggregates diminishes with the rise of temperature. With the increase of concentration the size of aggregates at 10 and 50° increases very little, whereas at 25° their size grows noticeably with the rise of concentration.

To determine whether the aggregates migrate onto the surface of the adsorbant, the structure formation was studied in the solutions before and after adsorption. It was shown that after adsorption in the solutions of oligomers the aggregates are not detected or else their dimensions are so small that the method employed does not permit their detection (the method of turbidity spectrum makes it possible to determine particle sizes up to 200 Å).

As is shown in fig. 2, in the epoxy resin adsorption larger aggregates migrate onto the surface of the adsorbent, leaving in the solution aggregates of smaller dimensions. In solutions of ЭД-5 and oligoethylene glycol adipinate the aggregates are not revealed right after the adsorption. The studies of the solutions after adsorption have shown that in a day or two, depending on solution concentration, the aggregates spring up again (see table 2), i.e., the system attains the state of equilibrium between the aggregated and free molecules.

Tables 3 and 4 present results of the studies of structure formation in two polyurethanes with molecular weights of 15000 and 60000 and in polycarbonate.

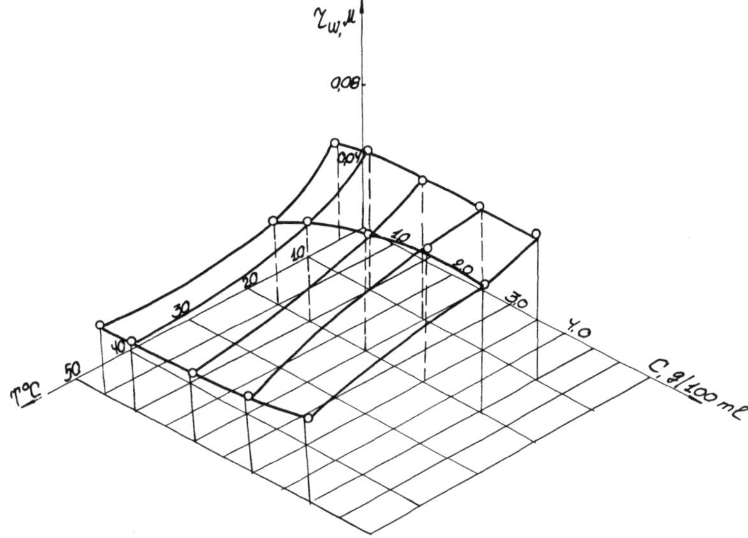

Fig. 4. Dependence of aggregate sizes of polycarbonate upon solution concentration and temperature (solvent dichlorethane)

Table 2. Change of aggregate sizes in solutions of oligoethylene glycol adipinate in time in solutions after adsorption

Solvent	Concentration C, g/100 ml	Time (hours)	Size $\bar{r}_w$, Å
Acetone	10.00	0	100
		4	100
		24	180
		48	300
		72	600
		336	1200
		360	1200
Toluene	3.00	0	100
		4	100
		24	300
		48	3200
		72	6000
		96	6000

From these data it is evident that the size of the aggregates of the polyurethanes varies with solution concentration. In solutions of low-molecular polyurethane right after adsorption within the range of concentrations from 1 to 3 g

Table 4. Structure formation in system polycarbonate-dichlorethane-Aerosil at 20 °C

Before adsorption			After adsorption	
C, g/100 ml	$\bar{r}_w$, mμ	$N \cdot 10^{-10}$	$\bar{r}_w$, mμ	$N \cdot 10^{-4}$
0.5	0.050	2.30	0.060	1.43
1.0	0.060	2.30	0.060	1.36
2.0	0.070	0.48	0.060	1.41
3.0	0.080	0.54	0.060	1.27
4.0	0.080	0.97	0.070	1.04

Table 3. Dependence of aggregate sizes upon molecular weight and concentration in system polyurethane-dimethyl formamide-glass spheres in solutions before and after adsorption, temperature 20 °C

	Before adsorption			After adsorption	
	C, g/100 ml	$\bar{r}_w$, mμ	$N \cdot 10^{-10}$	$\bar{r}_w$, mμ	$N \cdot 10^{-10}$
πy_1	0.5	0.030	1.80	not revealed	
	1.0	0.060	1.20	0.065	1.20
	2.0	0.070	0.20	0.070	0.20
	3.0	0.060	0.90	0.060	0.90
	4.0	0.040	1.30	not revealed	
	6.0	0.030	3.20	not revealed	
	8.0	0.030	2.80	ditto	
πy_2	0.8	0.080	84	0.072	$4.6 \cdot 10^{-3}$
	1.0	0.070	96	0.072	$2.8 \cdot 10^{-3}$
	2.0	0.050	217	0.057	$3.8 \cdot 10^{-3}$
	3.0	0.050	371	0.045	$4.1 \cdot 10^{-3}$

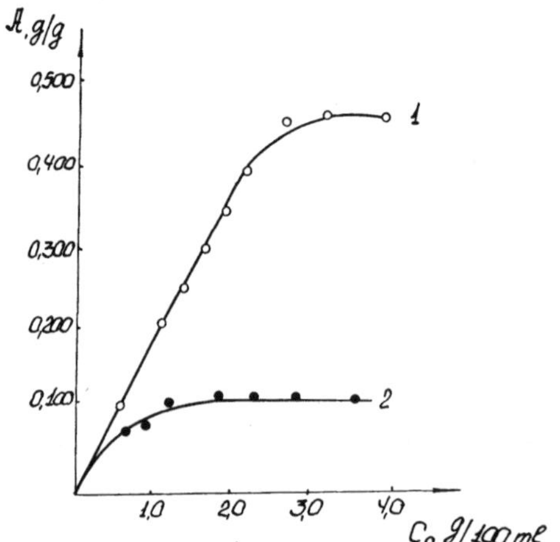

Fig. 5. Adsorption isotherms of oligoethylene glycol adipinate on Aerosil (1) and on carbon black (2) from solutions in toluene

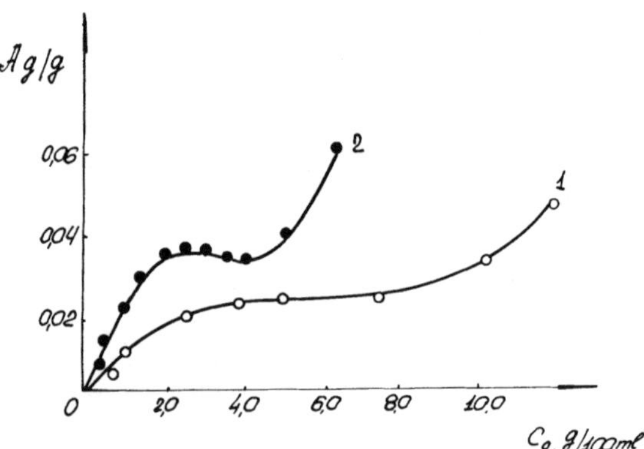

Fig. 6. Adsorption isotherms of oligoethylene glycol adipinate by glass spheres from solutions in dimethyl formamide (1) and acetone (2)

per 100 ml the sizes of aggregates do not vary, whereas in case of a polyurethane of a higher molecular weight and in polycarbonate the aggregates somewhat change their size with simultaneous reduction of their number, this indicating that in the course of the adsorption process the aggregates pass over to the surface. Thus, these data offer direct proof of the aggregates migrating onto the surface of the adsorbent during adsorption. Prevailing adsorption of aggregates as compared with non-aggregated macromolecules from the point of view of thermodynamics can be explained by the decrease of their solubility similar to the decrease of solubility of polymers with the increase of the molecular weight.

The transition of the aggregates onto the surface presupposes their weak bond with the surface, for only some of the molecules comprising the aggregate interact directly with the active centres of the surface. In fact, the experiments on desorption have proved that practically complete washing-out of the adsorbed polymer or oligomer takes place in the course of 15—30 minutes.

The transition of aggregates onto the surface of the adsorbent explains considerably greater values of the adsorption proper as compared to values of adsorption from diluted solutions. To illustrate that, some of the adsorption isotherms are given in figs. 5 and 6. Great adsorption values are explained by the transition of molecular aggregates onto the surface which considerably increases the number of linked molecules and layer thickness.

It was noted above that the size of the aggregates increases with the growth of solution concentration. The adsorption value rises correspondingly to a great extent at the same concentrations at which noticeable structure formation commences in the solution. Thus, the comparison of fig. 6 and 7 and table 2 shows that sharp increase of adsorption of oligoethylene glycol adipinate begins at concentrations in which sufficiently large aggregates are revealed. Here, in a number of cases the adsorption isotherms within the experimentally attainable concentration interval do not reach the state of saturation, whereas some isotherms with maximums are observed at a definite solution concentration (fig. 6). In some cases, e.g., in the adsorption of oligoethylene glycol adipinate from toluene, the isotherm has the usual appearance with saturation (fig. 5) typical for adsorptions from diluted solutions.

High adsorption values and the existence of aggregates in solutions make it possible to state that not only individual molecules migrate onto the surface but the aggregates as well, this is permitting the conclusion about "monoaggregated" coverage of the surface instead of monomolecular coverage. The size of the aggregates to a great degree determines the adsorption value. From the point of view of aggregating adsorption the saturation on isotherms can be determined by a great increase with concentration of the interaction between the aggregates and by the formation of a solid spatial net. As a result of this the transition of aggregates is either encumbered or stops with the rise of concentration (isotherms with maximums).

The investigation of dependence of epoxy resin adsorption upon the nature of the solvent made it possible to establish the phenomenon of inversion effect of the nature of solvent on adsorption within various concentration intervals (fig. 7). So, up to concentration 32—34 g per 100 ml the adsorption is higher from a poor solvent, — toluene, — while, with further rise of concentration the adsorption becomes higher from better solvents, such as acetone and dimethyl formamide.

To interpret these data we make use of data on structure formation in solutions (see fig. 1). Throughout the entire concentration range the size of aggregates in dimethyl formamide is smaller than in toluene. Correspondingly, the adsorption in the beginning is lesser from this solvent. But the increase of solution concentration in a poor solvent promotes the growth of aggregates and enhances interaction between them which leads to reduction of adsorption and to emergence of the isotherm on the plateau. In case of good solvents the degree of structurization is lower, the size of the aggregates changes little with the concentration and, as a result of this, the isotherm does not reach the saturation. All this leads to the observed inversion of adsorption dependence on the quality of the solvent.

Fig. 8 shows the isotherms of polystyrene adsorption from toluene (curve 1), cyclohexane at 34° (Θ-conditions, curve 2) and polycarbonate from chloroform (curve 3). The isotherms of polystyrene adsorption from toluene and polycarbonate from chloroform are of similar nature, they pass through maximum in the region of relatively diluted solutions and then noticeable rise of adsorption is observed which does not reach saturation. Experimental data on structure formation in polycarbonate solutions before and after adsorption (table 4) signify that the molecular aggregates are bonded with the surface. Rise of polystyrene adsorption higher than the definite solution concentration can also be associated with this reason, although we could not obtain direct experimental proof due to small differences in optical densities of solutions at various wavelengths which does not allow to employ the method of turbidity spectrum. Analysis of data on adsorption of polystyrene and polycarbonate allow to draw conclusions that the nature of adsorption isotherm changes considerably due to the change of thermodynamic properties of solvents and the

type of polymer linkage with the surface (the possibility of forming hydrogen linking of polycarbonate molecules with the surface of such adsorbent as glass). Shown in fig. 8 is isotherm of polystyrene adsorption obtained in a good solvent, — toluene, — and at point Θ (cyclohexane at 34°).

Evidently, the adsorption isotherm under Θ-conditions differs sharply in appearance from other isotherms and is step-like. Such difference in isotherm shapes is, in our opinion, associated with various changes of chain conformation and, consequently of the nature of structurization in

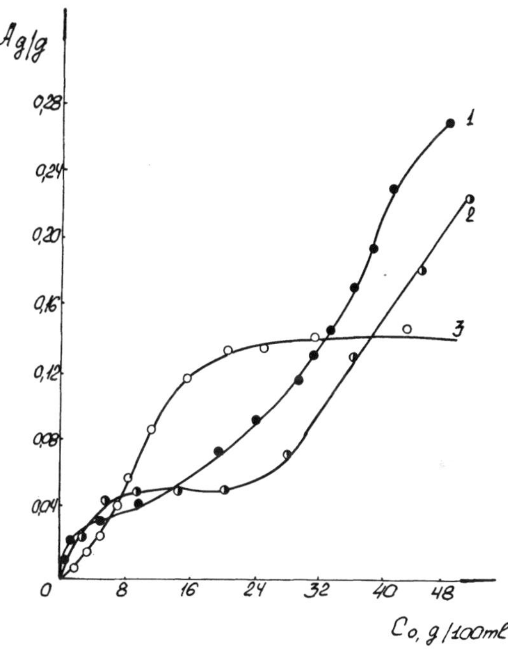

Fig. 7. Adsorption isotherms of epoxy resin ЭД-5 on glass spheres from solutions in dimethyl formamide (1), acetone (2) and toluene (3)

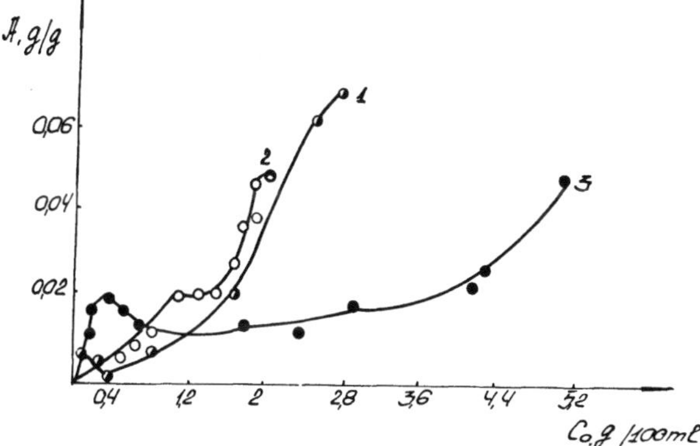

Fig. 8. Adsorption isotherms of polystyrene from toluene (1), cyclohexane (2) and polycarbonate from chloroform (3)

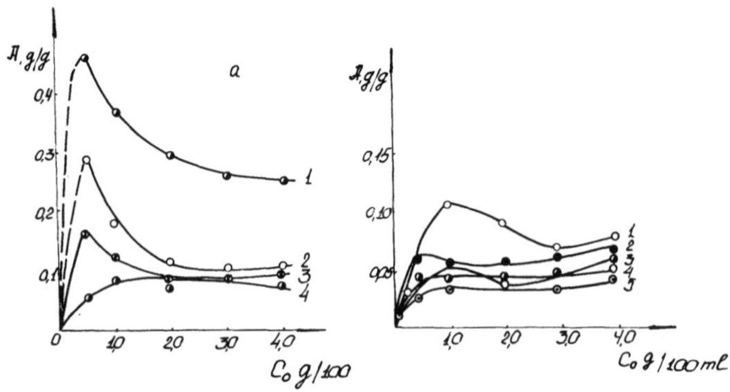

Fig. 9. Adsorption isotherms of polycarbonate (a) and polystyrene (b) at various concentrations of Aerosil at 25 °C: 1—5 mg SiO_2/ml; 2—10 mg/ml; 3—20 mg/ml; 4—30 mg/ml; 5—40 mg/ml

solution at concentration increase. It is well known that in Θ-solvent molecule sizes in diluted solutions are several times smaller than in a good solvent. Therefore, the adsorption value in diluted solutions is higher than in toluene (fig. 8). The increase of solution concentration leads to even greater contraction of the coils up to critical concentration of mixing (2) which is much higher than the concentration at which in a good solvent starts structure formation. As has been shown in a number of papers (2—5) at critical concentration the size of macromolecules is several-fold smaller than the size at point Θ even higher Θ-temperature. In addition to that, when the mixing concentration approaches the critical, the system develops great concentration fluctuations (5). Naturally, under such conditions the nature of aggregating will be other than in a good solvent. Whereas, in a good solvent the aggregate can be formed by mutually entangled chains, in Θ-solvent it consists, probably, of very contracted not entangled molecules. The polystyrene which was studied had a wide molecular-weight distribution. Therefore it can be assumed that in solutions of such polymers, as the concentration rises, first aggregates are formed as a result of interaction of chain coils of larger sizes, than from molecules of lesser molecular weight, etc. In other words, here some unusual "fractionating" of the aggregate seems to be possible. The appearance of steps on isotherms, when thus considered, can be explained by transition onto the adsorbent surface during concentration rise of various aggregate "fractions" consisting of molecules of different molecular weight.

In some investigations the dependence of adsorption value upon the quantity of adsorbent was shown experimentally in the system under conditions of equilibrium adsorption. We have conducted detailed studies of this phenomenon on systems polystyrene-CCl_4-Aerosil and polycarbonate-dichlorethane-Aerosil. Fig. 9 illustrates adsorption isotherms of both polymers at various contents of adsorbent in the system. As can be seen, with the increase of adsorbent quantity in the system the isotherms are disposed lower and lower as a result of the reduction of adsorption value. The maximum observed on the isotherms and conditioned by the commencement of structurization in solutions due to the increase of intermolecular interactions is smoothed out with the increase of adsorbent content. These dependences, it seems, can be explained by the specific adsorption from concentrated solutions, already considered above, in which molecular aggregates are adsorbed more readily than isolated macromolecules and where after adsorption a new state of equilibrium aggregate-molecule sets in the solution above the adsorbent. However, the results obtained (6—9) indicate that adsorption values are higher than, when all the adsorption would be conditioned only by the linking with the surface of the aggregates. It must be taken into account that simultaneously with the adsorption of aggregates the adsorption of isolated or non-aggregated molecules takes place. Since the total number of aggregates in the system is considerably lesser than the number of macromolecules in the solution, then, on the condition of complete linking of the aggregates (see table 2), it can be assumed that, the more the content of the adsorbent in the system, the greater part of the macromolecules has to be adsorbed in non-aggregated state, this leading to the reduction of total adsorption value. The disappearance of the maximum on the isotherms can be associated here with the change of conditions of structure formation when the adsorbent is added (destruction of aggregates similar to the disruption of the crystal

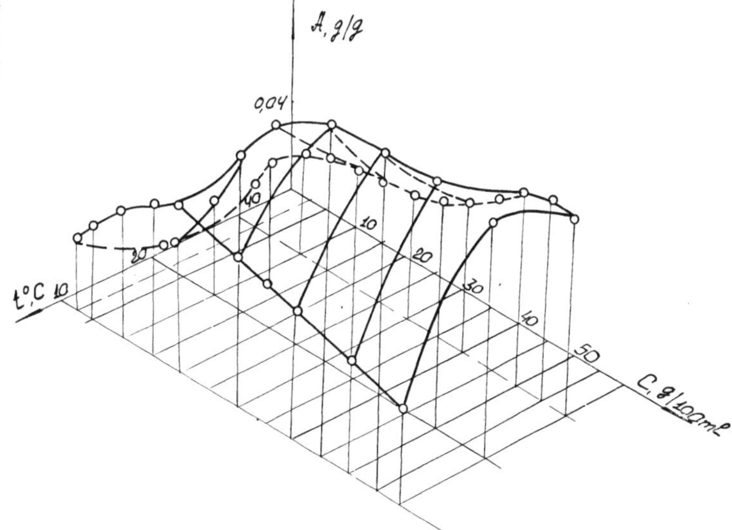

Fig. 10. Adsorption isotherms of epoxy resin ЭД-5 on glass spheres from solutions in toluene depending on temperature

structure of polymers when fillers are introduced into them). This can also be one of the causes of adsorption drop following the increase of the quantity of adsorbent.

Thus, we can draw an inference that the nature of dependence of polymer adsorption from concentrated solutions is extremely complex and is determined not only by conformational and structural changes in the solutions caused by the increase of solution concentration but by the influence of the adsorbent onto these processes. Adsorption aggregativity is one of the major distinctions of the adsorption from concentrated solutions as compared with the adsorption from diluted solutions or the adsorption of low-molecular substances. If in the latter case we deal with adsorbed particles of constant shape and composition, in the adsorption of polymers and oligomers with the change of concentration continuous changes of size and shape of the macromolecules take place along with the changes of the degree of aggregation and the nature of structure formation. Actually at each concentration we have an other structure of the sorbed particles. This explains the complex dependence of adsorption upon the concentration of polymer solution.

The value and nature of isotherms of adsorption from concentrated solutions are greatly effected by temperature. Its influence is manifested first of all in the dependence of the nature of processes of structure formation upon temperature. Taking this concepts into account, let us consider the effect of temperature upon the

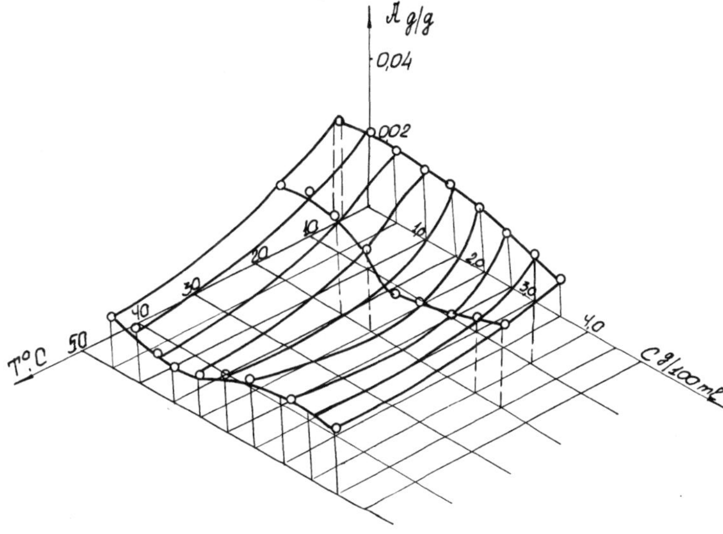

Fig. 11. Adsorption isotherms of polycarbonate from solutions in dichlorethane on Aerosil depending on temperature

adsorption of epoxy resin and polycarbonate (figs. 10 and 11). The analysis of obtained results indicates that the rise of temperature leads not only to the change of adsorption value but also to the change of the nature of the isotherms themselves. So, at 10° the maximum is observed on the adsorption isotherm, at 25° the isotherm reaches saturation after the concentration of 20 g/100 ml; at 40° the adsorption drops, while the isotherm does not reach saturation. The data on polycarbonate (fig. 11) conform this general pattern. The obtained results can be well explained on the basis of data about temperature dependence of the structure formation. The rise of temperature leads to destruction of aggregates in diluted solutions and to the reduction of aggregate sizes in concentrated solutions (fig. 3). In fact, with the rise of temperature the start of the structure formation shifts toward the high concentrations, the size of aggregates gets smaller, whereas the concentrational dependence of aggregate sizes becomes less distinct with the rise of temperature. These data make it possible to explain the phenomenon of adsorption reduction with the temperature by the reduction of the degree of structure formation in solutions. Thus, the comparison of data on adsorption and structure formation with temperature changes confirms the presence of direct connection between the adsorption capacity of macromolecules and structure formation in solution.

These processes can be qualitatively described on the basis of a simplest model of aggregate formation. According to (10) in polymeric solutions the existence of equilibrium between the aggregated and non-aggregated macromolecules should be admitted:

$$\pi M \rightleftarrows \pi^{\mathrm{I}} A + \pi^{\mathrm{II}} M,$$

where: πM total number of macromolecules in a unit of volume;

$\pi^{\mathrm{I}} A$ number of molecules linked into aggregates;

$\pi^{\mathrm{II}} M$ number of free molecules.

Hence, the aggregation constant can be expressed as the relation of the linked molecules to the total number of molecules in the system:

$$K_a = \pi^{\mathrm{I}}/\pi .$$

Presenting the aggregate of molecules in our paper (9) like a result of association of coiled molecules and like a spherical formation, with the average size of the aggregates determined experimentally taken into account, calculations of aggregation constants were carried out (table 5). The dependence of aggregation constants for epoxy resin upon temperature and concentration is shown in fig. 12. Attention is attracted by the fact that, with the increase of solution concentration, the aggregation constant decreases, while the effect of temperature can lead to the appearance of maximum constant

Table 5. Dependence of aggregation constant K_a upon concentration and temperature for epoxy resin in toluene

Temp. °C	Concentration C, g/100 ml	Aggregate size $\bar{r}_w$, mμ	Quantity $N \cdot 10^{-9}$, cm^{-3}	$K_a \cdot 10^5$
10	10.0	0.11	9.00	4.00
10	15.0	0.14	0.80	0.49
10	20.0	0.16	0.40	0.27
10	30.0	0.20	0.09	0.08
10	40.0	0.23	0.09	0.09
10	50.0	0.25	0.09	0.09
10	60.0	0.27	0.09	0.09
25	15.0	0.06	9.90	0.47
25	20.0	0.09	3.40	0.41
25	30.0	0.12	2.80	0.53
25	40.0	0.14	1.70	0.38
25	50.0	0.15	1.30	0.30
25	60.0	0.16	1.20	0.27
40	20.0	0.04	20.0	0.21
40	30.0	0.05	22.0	0.30
40	40.0	0.05	24.0	0.33
40	50.0	0.05	26.0	0.29
40	60.0	0.05	27.0	0.25

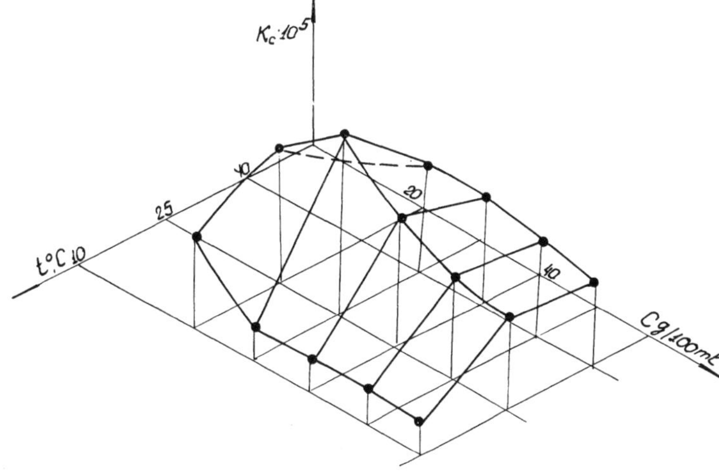

Fig. 12. Dependence of aggregation constants (K_a) upon solution concentration and temperature in system epoxy resin-toluene-glass spheres

value at certain conditions. It is significant that maximum adsorption values correspond to solutions in which the value of aggregation constant attains the maximum (fig. 10). Thus, having considered the adsorption dependence upon the temperature, the conclusion can be drawn that its value substantially depends upon the value of aggregation constant and upon its temperature dependence. The change of aggregate sizes with the change of temperature and the degree of interaggregate interaction determine the change of adsorption with change of temperature, leading to the appearance of complex dependences. At the same time, with the presence of aggregating adsorption mechanism, it has to be borne in mind that this mechanism coexists with adsorption of non-aggregated macromolecules. The existence of the equilibrium aggregate-molecule in the solution and low values of aggregation constants indicate that during adsorption a shift of the equilibrium takes place in the solution as a result of preferable (but not exclusive) adsorption of aggregates. The above experimental data indicate that adsorption values are higher than if it would be determined only by the transition of the aggregates onto the surface. Actually, with the rise of solution concentration, the total number of macromolecules in it increases considerably faster than the number of aggregates (see fig. 13) and, correspondingly, the total degree of aggregation drops. Hence, the adsorption of aggregates is superimposed by the usual adsorption of isolated macromolecules whose conformation, however, differs from the same in a diluted solution due to the effect of the overlapping of coils. As was already noted, this is one of the

reasons for the appearance of dependence of the equilibrium adsorption value upon the quantity of the adsorbent introduced into the system. This concept can be confirmed by data obtained in the investigation of molecular mobility of adsorbed chains in the adsorption layer estimated by the method of nuclear magnetic resonance. In contrast to our other work, we conducted investigations under conditions of equilibrium of the adsorbed polymer with a polymer in solution, i.e., without separating the adsorbent from the equilibrium polymer solution. It has been established that, with the increase of the content of the adsorbent in the system, the relative intensity of signals decreases (fig. 14), i.e., the molecular mobility in the adsorption layer becomes lesser. As it was already mentioned above, under such conditions the total adsorption value drops as well. In view of simultaneous adsorption of both of the aggregates and separate macromolecules the reduction of mobility due to the increased content of the adsorbent can be explained by the fact that in

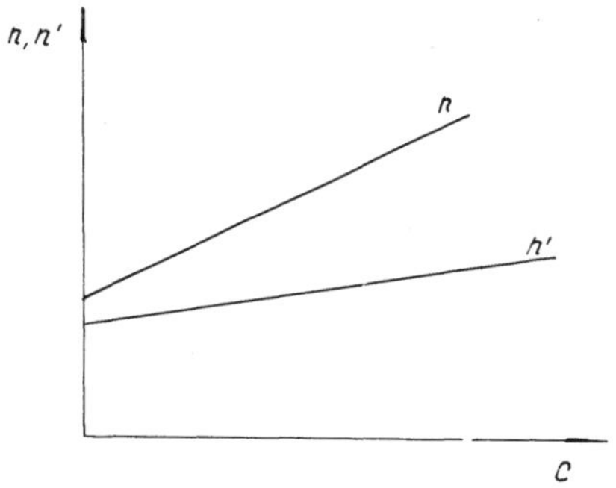

Fig. 13. Dependence of total quantity of molecules in solution (n) and number of molecules linked into aggregates (n') upon solution concentration

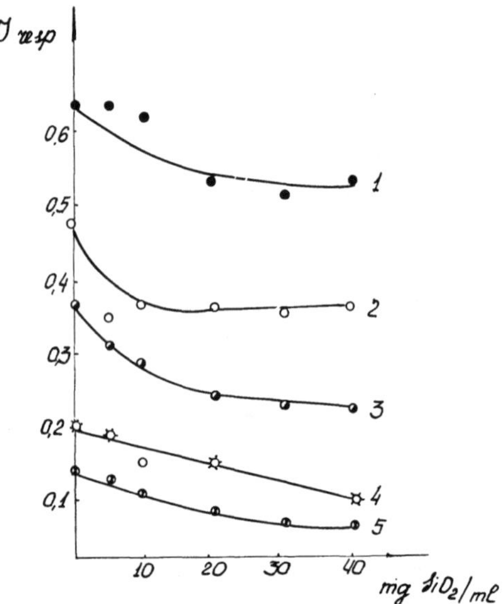

Fig. 14. Dependence $I_{rel.}$ of signal of meta- and para-protons (1—5) of a benzine ring for polystyrene solutions upon the quantity of Aerosil at various solution concentrations: 1—4.0 g/100 ml; 2—3.0 g/100 ml; 3—2.0 g/100 ml; 4—1.0 g/100 ml; 5—0.5 g/100 ml

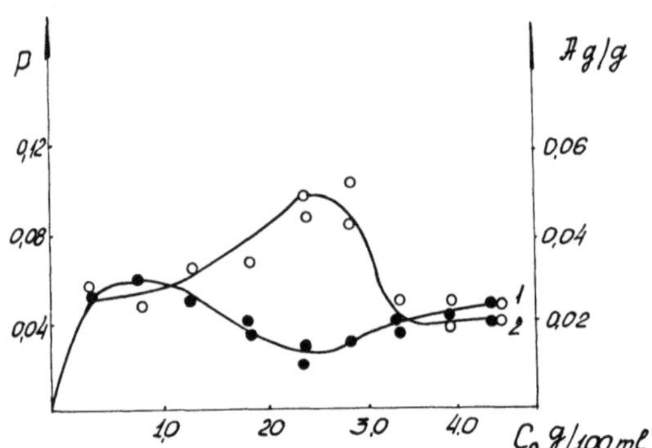

Fig. 15. Adsorption isotherm (1) and dependence of the part of linked segments upon solution concentration (2) in system polycarbonate-dichlorethane-Aerosil at 20 °C

the adsorption layers formed at high contents of the adsorbent the portion of aggregates decreases and the part of non-associated macromolecules increases. As a result of this, a greater total reduction of mobility takes place in such a layer than in layers formed by aggregates only, and where the larger part of the molecules comprising the aggregate does not interact with the surface directly. All the above concepts should lead to an inference that, in contrast to the adsorption in diluted solutions, in case of concentrated solutions the part of segments linked directly with the surface must be less. In fact, as a result of the macromolecules being a part of the composition of aggregates, considerably lesser part of the adsorbed molecules has a possibility of direct interaction with the surface. In our papers (11), (12) we determined the part of linked segments in the adsorption layer obtained from concentrated solution and compared it with adsorption isotherms. The study was carried out for the system polycarbonate-Aerosil and oligoglycol adipinate-Aerosil. Represented in figs. 15—17 are adsorption isotherms of the investigated polymers and dependences of the part of linked segments determined by the well known for this purpose method of infrared spectroscopy upon solution concentration. Our attention was attracted by really small values of the part of linked segments "p" — in the system polycarbonate-Aerosil values "p" are 0.04—0.08, i.e. by an order lower than was usually observed in diluted solutions. On the other hand, dependence "p" upon concentration is antibate to the dependence of adsorption value upon solution concentration. As a result of this maximum is observed on the

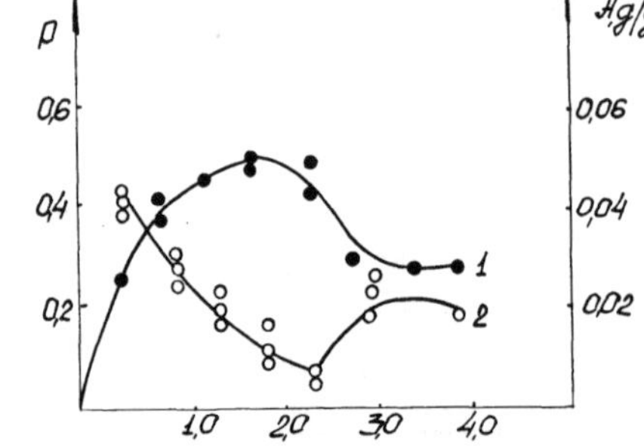

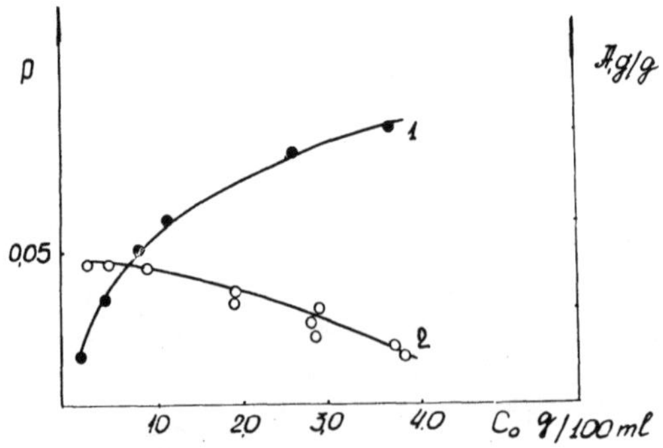

Fig. 16. Adsorption isotherm (1) and dependence of the part of linked segments upon solution concentration (2) in system oligoethylene glycol adipinate-dichlorethane-Aerosil at 20 °C

Fig. 17. Adsorption isotherm (1) and dependence of the part of linked segments upon solution concentration (2) in system oligoethylene glycol adipinate-toluene-Aerosil at 20 °C

"p"-C dependence curve in those regions where minimums exist on the adsorption isotherms. This means the lesser the adsorption value, the greater the part of linked segments. This result confirms the above concepts that molecule aggregates pass over onto adsorbent surface, the majority of these molecules is not linked with the surface. When the adsorption value rises on account of aggregate adsorption, the degree of macromolecule linkage by the surface decreases, while the layer thickness increases. The external course of dependence "p" upon concentration beyond the point of extremum can be associated with the change of shape and orientation of aggregates in the surface layers leading to the change of conditions of their interaction with the surface and, correspondingly, to the change of value "p" (e.g. growth of "p" in the concentration region 2.0—3.0 g/100 ml for polycarbonate at the drop of total adsorption). Thus, in contrast to the adsorption from diluted solutions in which value "p" always decreases steadily with the rise of solution concentration, in the adsorption from concentrated solutions, as can be seen from the presented data, value "p" can change non-monotonically with the rise of solution concentration, the same applies also to the adsorption value. In addition to that, value "p" is effected by the quality of the solvent as a factor influencing the structure formation. So, in case of the system oligoethylene glycol adipinate-toluene or dichlorethane was shown that the size of aggregates in a poor solvent is larger and, correspondingly, the adsorption value is higher, while the part of linked segments is lesser (see figs. 16 and 17).

In this way, the degree of linking of macromolecules with the surface in concentrated solutions is also associated with solution structure determined by the concentration and by thermodynamic quality of the solvent.

The stated data show also that the structure and properties of adsorption layers formed from solutions of various concentrations will also be determined by conditions of structure formation in solution. We have a great amount of experimental data at our disposal confirming this idea. Continuous change of the nature of structure formation in solution substantially affects the properties of the adsorption layers formed both by the aggregates and by isolated macromolecules. It must be pointed out that, according to the above statement, the properties of the layers change along with the change of their thickness in an equally complicated way as is the case with the adsorption itself.

In conclusion the author expresses sincere gratitude to Dr. *L. M. Sergeyeva* who took part in all the work on which this article is based.

References

1) *Lipativ, Yu. S.* and *L. M. Sergeyeva*, Adsorption of Polymers (New York 1974).
2) *Debye, P., D. Woermann,* and *B. Chu,* J. Chem. Physics **36**, 7 (1962).
3) *Eskin, V. E.* and *A. E. Nesterov,* Vysokomolecularniye Soyedineniya 8, 1045 (1966).
4) *Eskin, V. E.* and *A. E. Nesterov,* Kolloidniy Journal **6** 132 (1966).
5) *Debye, P.,* J. Chem. Physics **31**, 680 (1959).
6) *Lipatov, Yu. S., T. T. Todosijchuk,* and *L. M. Sergeyeva,* DAN Ukr. SSR, Seriya B, No. 5 (1971).
7) *Lipatov, Yu. S., T. T. Todosijchuk,* and *L. M. Sergeyeva,* Vysokomolecularniye Soyedineniya **B 14**, 121 (1972).
8) *Sergeyeva, L. M., N. A. Chesnokova, T. T. Todosijchuk,* and *E. I. Zabarovskaya,* V sb, Heterogenniye Polymerniye Materialy. "Naukova Dumka", K., str. 96 (1973).
9) *Lipatov, Yu. S., T. T. Todosijchuk,* and *L. M. Sergeyeva,* V sb, Syntez i Physicokhimiya Polymerov. "Naukova Dumka", K., vyp. 13, str. 79 (1974).
10) *Lipatov, S. M.,* Vysokomolecularniye Soyedineniya. Izd. AN BSSR (1943).
11) *Lipatov, Yu. S., L. M. Sergeyeva, T. T. Todosijchuk,* and *T. S. Khramova,* Kolloidniy Journal **37**, 280 —284 (1975).
12) *Lipatov, Yu. S., L. M. Sergeyeva, T. T. Todosijchuk,* and *T. S. Khramova,* DAN SSSR **218**, 1144—1146 (1974).

Author's address:

Yu. S. Lipatov
The Institute of Macromolecular Chemistry
at the Academy of Sciences
of the Ukrainian SSR
Kiev 160 (USSR)

Progr. Colloid & Polymer Sci. **61**, 24–35 (1976)
© 1976 by Dr. Dietrich Steinkopff Verlag GmbH & Co. KG, Darmstadt
ISSN 0340-255 X

Plenary lecture of the IUPAC-Conference on Colloid and Surface Science in
Budapest, September 15–20, 1975

*Institute of Colloid and Surface Science and Department of Chemistry,
Clarkson College of Technology, Potsdam, New York (USA)*

Preparation and characterization of monodispersed metal hydrous oxide sols *)

E. Matijević

With 20 figures

(Received December 20, 1975)

Introduction

Over the past few years considerable theoretical
advances have been made in developing an
understanding of the problems of colloid stability.
In order to test the various models, it is essential
to have colloidal dispersions well defined in
terms of particle size, shape, charge, and other
characteristics. Until recently, essentially the
only such systems have been the monodispersed
organic latices consisting of spherical particles
in micron and submicron sizes.

We have now produced an entirely new
family of inorganic colloidal dispersions with
particles of exceedingly narrow size distribution
and of varying shapes, which include spheres,
cubes, cube-octahedra, needles, ellipsoids, etc.
Chemically, all these systems are metal (hydrous)
oxides.

The novel sols are excellent models for the
study of colloid stability, since their surfaces are
rather well-defined and because their charge can
be conveniently changed in magnitude and in
sign just by varying the pH. The inorganic
latices to be described are also of interest in
various applications, such as pigments, catalysts
and catalyst carriers, coatings, fillers, etc.

In view of the importance of metal hydroxides
it is no surprise that countless studies have been
reported on their precipitation and characteriza-
tions. Yet, there is no good theory on the
mechanisms of formation and of the growth of
such systems. This is easily understood if one
considers that generation of metal hydroxides

depends on a large number of factors, the most
important being pH, temperature, aging, and
the nature of the anions, in addition to the
parameters which affect any solid phase formation (e.g., the concentrations of the reacting
components, method of mixing, etc.).

The rather dramatic role exhibited by the pH
and by the anions in solution indicates that
specific solute metal complexes precede or even
initiate nucleation of metal hydrous oxides. The
availability of monodispersed inorganic latices,
which can be reproducibly prepared, has now
made it possible to elucidate the reactions which
lead to the solid phase formation and to study
the mechanism of particle growth. The understanding of these processes is of interest in many
areas, but probably it is of greatest importance
in the explanation of corrosion of metals in
general, and of iron in particular.

In this paper we will describe the methods of
preparation of a number of monodispersed metal
hydrous oxide sols including those of chromium,
aluminum, iron, titanium, thorium, and copper.
Furthermore, the effects of anions and of other
additives on particle size, shape, and composition
will be illustrated. Finally, at least in the case of
chromium, the mechanism of formation of its
hydroxide will be discussed.

Method of preparation of monodispersed
metals hydrous oxide sols

The essential aspect of the procedures for the
preparation of metal hydroxide colloidal dispersions consists of forced hydrolysis of metal
ions in acidic (sometimes very highly acidic)
aqueous solutions of metal salts by aging the

---

*) The work was supported by the NSF Grant GP
42331X and by the American Iron and Steel Institute
project No. 63-269.

latter at elevated temperatures (75—180 °C) for varying periods of time (20 minutes to several weeks). Simultaneously, the nature and the concentration of the anions have to be carefully controlled, as these exercise a dominant effect on the composition, size, and shape of the final sol particles. It is noteworthy that sulfate and phosphate ions seem to play the most important role in the formation of monodispersed metal hydrous oxide sols. For example, only in the presence of these anions have spherical particles of chromium, aluminum, thorium, and zirconium hydroxides been generated. In their absence either sols do not appear (chromium salt solutions) or particles of entirely different shapes (rods, ellipsoids, etc.) precipitate on aging (aluminum or ferric salt solutions).

The optimal temperatures, times of aging, and the concentration of metal ions and anions vary from case to case. For example, ∼ 24 hours at 75 °C are needed to obtain a monodispersed chromium hydroxide sols, whereas only 2 hours at 98 °C will give an excellent ferric basic sulfate dispersion. The rate of heating also plays a role, but mostly with regard to the degree of uniformity.

In some instances, the hydrolysis may be assisted by homogeneous generation of hydroxyl ions, such as happens when the aging is carried out with urea added to the metal salt solutions.

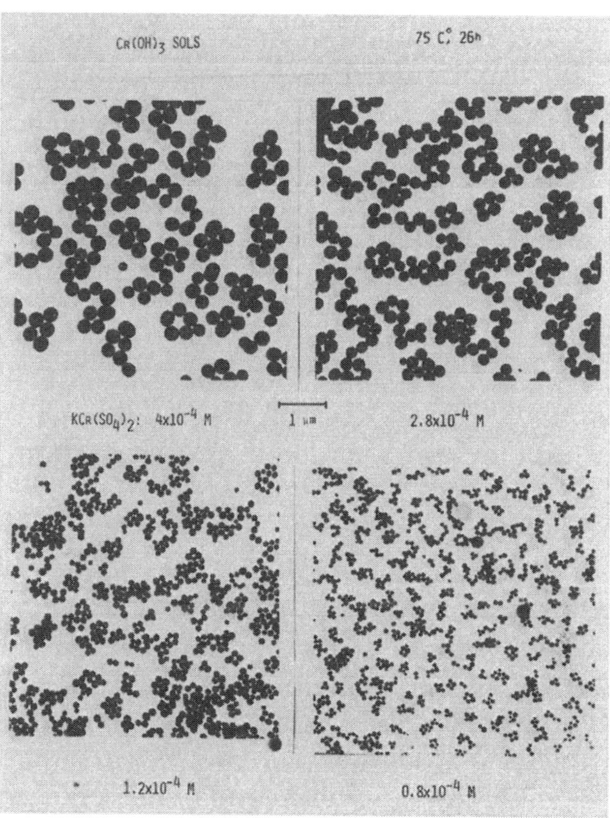

Fig. 1. Electron micrographs of chromium hydroxide sol particles obtained by heating at 75 °C for 26 hours solutions of chrome alum of four different concentrations

Examples of monodispersed metal hydrous oxide particles

By using the described procedure and by properly adjusting the experimental parameters a number of metal hydrous oxides of narrow size distribution have been produced. Noteworthy, depending on the anions present in the aged solutions, some metals will give either spherical (amorphous) or crystalline particles. Hydrosols consisting of spherical particles are especially useful as they can be readily analyzed in situ for size distribution and number concentration by light scattering. This information is pertinent in the study of particle nucleation and growth mechanisms.

a) Chromium

Fig. 1 gives four electron micrographs of chromium hydroxide sols obtained by aging for 26 hours at 75 °C, chrome alum solutions of different concentrations. In all cases particles are spherical and of narrow size distributions; their modal radius decreases as the concentration of the solution becomes lower (1). Similar results are obtained if chromium nitrate, chloride, or perchlorate are heated in the presence of a soluble sulfate salt, but not in the absence of sulfate ions; in the latter case these chromium salts solutions produced no sols on aging. The particle size also depends on the chromium to sulfate ratio; with increasing sulfate concentration the same chromium salt solutions yield larger particles (1, 2). The addition of phosphate ions has a similar effect as sulfate ions on the formation of chromium hydroxide sols (3).

It is important to recognize that the spherical particles are amorphous and that they contain no sulfate (4), although the latter species is essential to the solid phase formation. The role of sulfate ions will be discussed below.

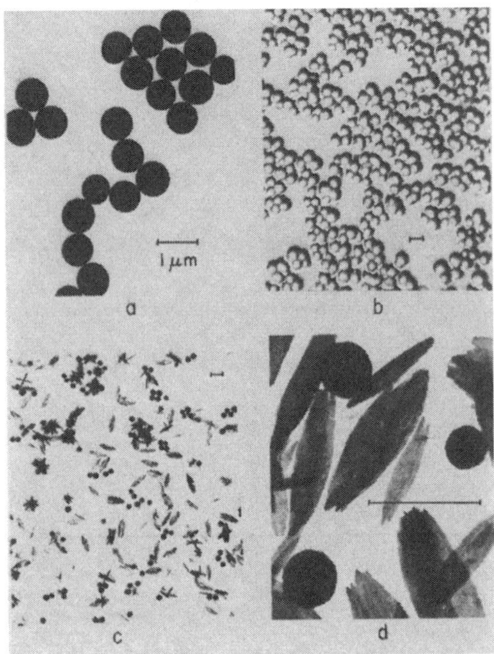

Fig. 2 (a). Electron micrograph of aluminum hydrous oxide sol particles prepared by aging a 2×10^{-3} M solution of $Al_2(SO_4)_3$ at 98 °C for 84 hours. (b) Electron micrograph of a carbon replica of aluminum hydrous oxide particles obtained by aging 1×10^{-3} M solution of $KAl(SO_4)_2$ at 98 °C for 24 hours. (c) Electron micrograph of a sol prepared by aging of a 1×10^{-3} M solution of $Al_2(SO_4)_3$ for 24 hours at 98 °C followed by adjustment of pH to 9.7 by NaOH, deionization, repeptization at pH $\sim$ 8, and additional aging at 98 °C for 24 hours. (d) The same as (c) but at higher magnification. The lines indicate 1 μm in all cases

b) Aluminum

Fig. 2 shows a transmission electron micrograph (a) and a replica (b) of an aluminum hydrous oxide sol obtained by aging an $Al_2(SO_4)_3$ solution (5). Again, spherical particles form, but only in the presence of sulfate or phosphate ions. Aluminum chloride, nitrate, or perchlorate solutions yield on aging anisometric particles (ellipsoids, rods, etc.). Unlike chromium hydroxide, the particles of aluminum hydrous oxides prepared from aluminum sulfate solutions contain appreciable amounts of sulfate ions, but these seem to be weakly bound and are readily exchangeable. On heating the deionized sol with dilute base (at pH 9.7) the entire content of sulfate in the solids is substituted by hydroxyl ions. As a consequence, the particles shrink a little but retain their spherical shape. Fig. 3 gives as an example the size distribution curves

for two different sols as obtained by light scattering before and after sulfate was exchanged from the aluminum hydrous oxide particles.

Once the sulfate ions are removed, further aging of the sols brings about crystallization of the particles and consequently a change in shape. Fig. 2 (c and d) shows such a transition. Originally all particles were perfectly spherical. and if the secondary aging had been continued they would have all appeared as crystalline platelets. The change in the structure is illustrated in the electron diffraction patterns given in fig. 4 for the amorphous spherical particles and in fig. 5 for crystallized material obtained as described above. The latter corresponds to bohemite.

c) Thorium

Most recently, sols consisting of spherical particles of thorium hydrous oxide were prepared by aging of thorium nitrate solutions in the presence of Na_2SO_4. One example given in

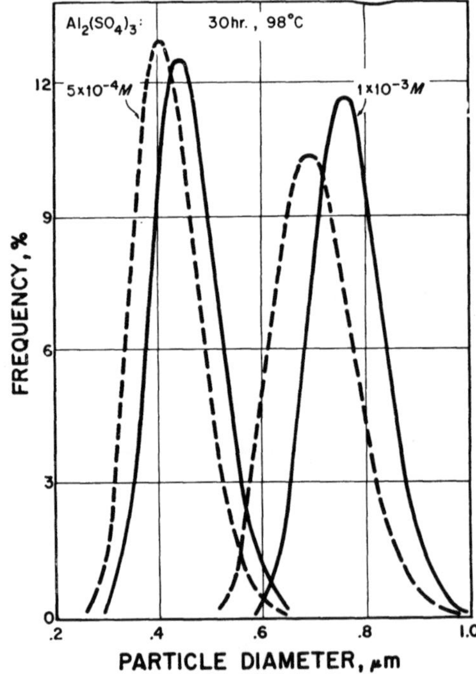

Fig. 3. Particle size distribution curves as obtained by light scattering for two different aluminum hydrous oxide sols prepared by aging at 98 °C for 30 hours 5×10^{-4} M and 1×10^{-3} M solutions of $Al_2(SO_4)_3$, respectively (solid lines). Dashed lines give the corresponding size distributions of the same sols after all sulfate ions in the particles had been exchanged for hydroxyl groups

Fig. 4. Electron diffraction pattern of spherical particles of an aluminum hydrous oxide sol obtained by aging an $Al_2(SO_4)_3$ solution (corresponding to sols as illustrated in fig. 2 (a and b). The particles contain sulfate ions

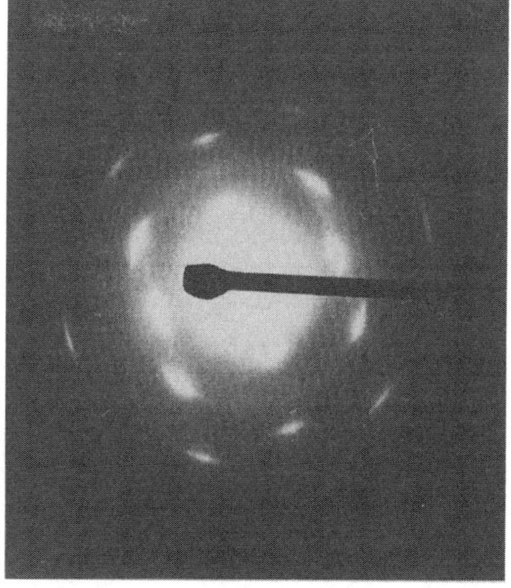

Fig. 5. Electron diffraction pattern of an aluminum hydrous oxide sol consisting of platelets corresponding to sols illustrated in fig. 2 (c and d), obtained by re-aging of spherical aluminum hydrous oxide particles after exchanging all sulfate ions for hydroxyl groups

Fig. 6 shows that size of the particles is not as uniform as of the system shown earlier; but there is no doubt that conditions may be refined so that eventually sols of much narrower size distributions will be obtained. Significantly, no particles appear in the aged solutions in the absence of sulfate ions.

d) Iron

Of different metals, ferric ions are among the most reactive. Thus, the feasibility of the preparation of colloidal monodispersed ferric hydrous oxides seemed rather remote. However, we succeeded in establishing conditions which gave such particles of excellent uniformity in a variety of shapes. Again the composition of the particles and their morphology was strictly dependent on the nature and the concentration of the anions present in the solutions that were aged.

Fig. 7 is an electron micrograph of a colloidal ferric basic sulfate obtained by aging a solution containing $Fe(NO_3)_3$ and Na_2SO_4. Although the particles appear spherical, they are actually of hexagonal symmetry as clearly seen from the replication of the same sample (fig. 8); the scanning electron micrograph (insert, fig. 8) nicely confirms the morphology. When the ratio of ferric to sulfate ions is altered, particle shape

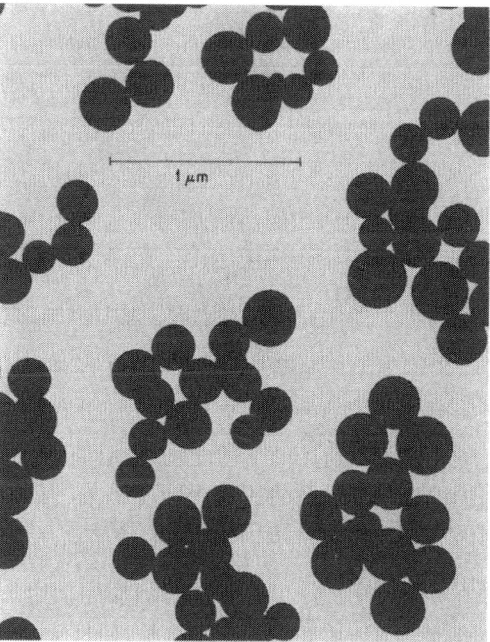

Fig. 6. Electron micrograph of thorium hydrous oxide sol particles obtained by aging for 2 hours at 60 °C a solution 5×10^{-4} M in $Th(NO_3)_4$ and 5×10^{-4} M in Na_2SO_4 (initial pH $= 3.35$, final pH $= 3.22$)

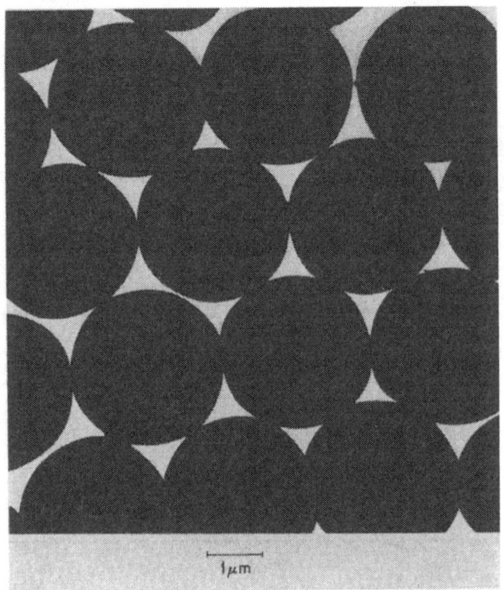

Fig. 7. Electron micrograph of basic iron(III) sulfate sol particles prepared from a solution 0.18 M in Fe(NO$_3$)$_3$ and 0.53 M Na$_2$SO$_4$ aged in an oil both heated from room temperature to 80 °C at a rate of 1.5 °C/min and then maintained at 80 °C for 1.5 hours

also changes as illustrated in fig. 9; the replica of the same system is given in fig. 10. Two kinds of particles are indicated: hexagonal and monoclinic and their structures were confirmed by X-ray analysis. The chemical composition of

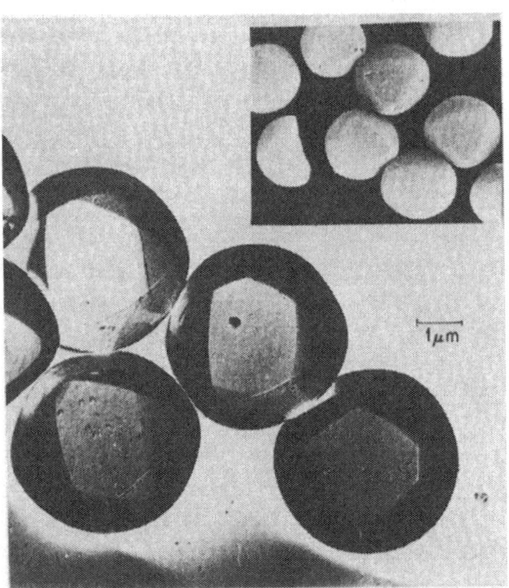

Fig. 8. Electron micrograph of a carbon replica of basic iron(III) sulfate particles similar to those described in fig. 7. Insert shows a scanning electron micrograph of the same particles

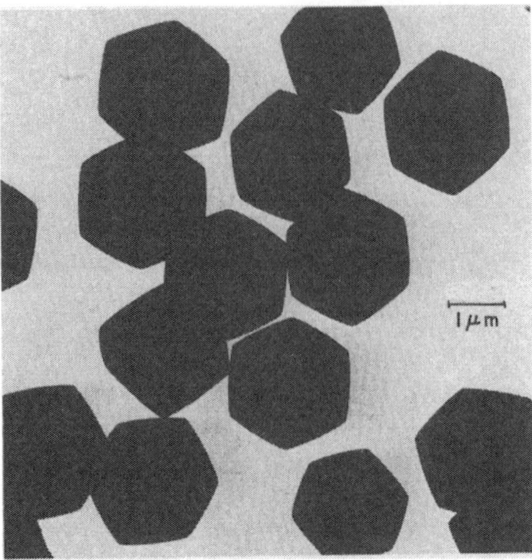

Fig. 9. Electron micrograph of basic iron(III) sulfate particles prepared by aging a 0.088 M solution of Fe$_2$(SO$_4$)$_3$ in an oven at 98 $\pm$ 5 °C for 2.8 hours

Fig. 10. Electron micrograph of a carbon replica of the particles described in fig. 9

these particles was identified as Fe$_3$(OH)$_5$(SO$_4$)$_2$ · 2 H$_2$O (hexagonal) and Fe$_4$(OH)$_{10}$(SO$_4$) (monoclinic), the former corresponding to the alumite group minerals (6).

The anions again have a great effect on the formation of ferric hydrous oxide sols. Solutions of Fe(NO$_3$)$_3$ yielded ellipsoidal particles (fig. 11) whereas FeCl$_3$ solutions on heating gave rod-like particles (fig. 12). X-ray analysis showed the former to correspond to α-Fe$_2$O$_3$, and the latter to β-FeO(OH).

The presence of cations other than those of the alkali metals also influenced the nature of the sols formed. This was manifested essentially in the change in particle size as illustrated in fig. 13 in which the modal diameter is plotted against the mole fraction of divalent cations used to substitute sodium ions in the system $Fe(NO_3)_3$—Na_2SO_4. In all cases the concentration of SO_4^{2-} was kept constant in order not to affect particle composition and shape by changing the content of sulfate ions in the growth medium. The complete inhibition of particle formation when MSO_4 ($M = Cu^{2+}$, Ni^{2+}, Mg^{2+}, or Cd^{2+}) was used alone (mole fraction of $MSO_4 = 1$) was found to be due to the change in pH. As the fractional concentration of MSO_4 was increased the pH became lower; for example, in the case of gradual substitution of Na_2SO_4 with $NiSO_4$, the pH dropped from 1.78 (mole fraction of $NiSO_4 = 0$) to pH = 1.45 (mole fraction of $NiSO_4 = 1$). If the same experiments were repeated at constant pH (1.77—1.80), the results shown in fig. 14 were obtained. Apparently the decrease in particle size (and eventually the inhibition of particle formation at still higher $NiSO_4$ concentrations) was caused by the acidification of the solutions resulting from the addition of the second electrolyte. These data point out the extreme sensitivity to pH of the processes involved in the formation of ferric hydrous oxides. Obviously, the change in pH affects the complexes in solution which act as the precursors to precipitation. The nature of these complexes at various temperatures is now being investigated.

e) Copper

Monodispersed sols were also obtained by aging copper salt solutions (similar to Fehling's solutions, but of different concentrations of components) in the presence of glucose (7). All particles were of cubic crystal symmetry characteristic of copper(I) oxide, but their habitus differed depending on the concentration of the reacting components. Fig. 15 gives schematically various particle shapes as they develop in transformation from cubes to octahedra. Fig. 16 is a reproduction of four electron microscope replicas of copper(I) oxide systems showing that indeed all these shapes can be obtained.

Fig. 11. Electron micrograph of ferric hydrous oxide sol particles prepared by aging a solution 2.4 M in $Fe(NO_3)_3$ and 1.9 M in HNO_3 for 14 minutes in a silicone oil both at $180 \pm 1\,°C$

f) Titanium

The size and shape of titanium dioxide particles are of particular importance when this material is used as pigment. By extensive aging of titanium chloride solutions in the presence of sulfate ions, it was possible to produce uniform titanium dioxide sols consisting of spherical

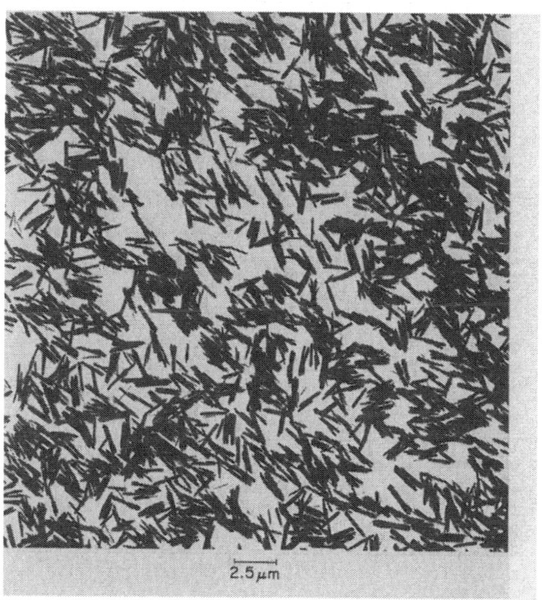

Fig. 12. Electron micrograph of ferric hydrous oxide sol particles prepared by aging a solution 0.36 M in $FeCl_3$ and 0.001 M in HCl in an oven at $110\,°C$ for 24 hours

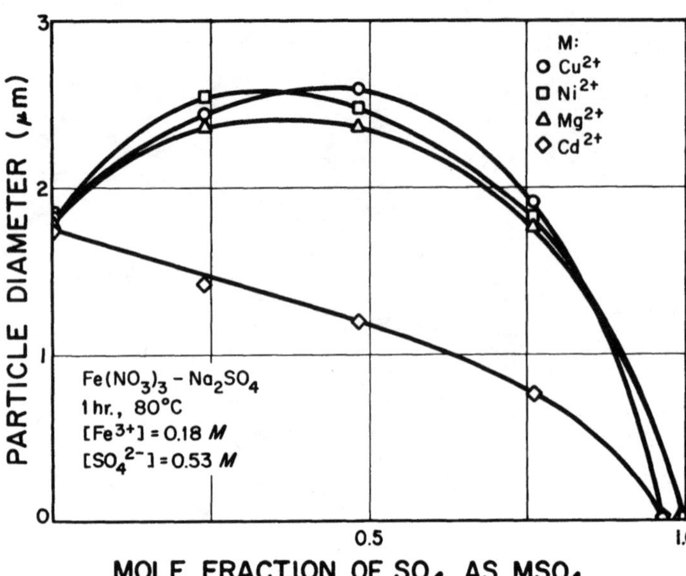

Fig. 13. Diameter of basic iron(III) sulfate particles as a function of the mole fraction of $CuSO_4$($\bigcirc$), $NiSO_4$($\square$), $MgSO_4$($\triangle$), and $CdSO_4$($\diamond$), respectively. The sols were obtained by heating solutions 0.18 M in $Fe(NO_3)_3$ and 0.53 M in SO_4^{2-} at 80 °C for 1 hour. The balance of sulfate was added as Na_2SO_4

particles. Their size could be altered by the length of aging. Scanning electron micrograph (fig. 17) illustrates such a sol with particles of 4.0 μm diameter, the uniformity of which is rather striking. Presently, the somewhat puzzling aspect of this system is that despite the apparently spherical habitus, the particles exhibit a definite X-ray pattern characteristic of rutile. This latter composition is further substantiated by the finding that the refractive index which best fits the light scattering data taken with the same sol is that of rutile (8).

Studies with monodispersed metal hydrous oxide sols

The availability of inorganic latices described in the previous section has made it possible to carry out a number of investigations on various aspects of the properties of colloidal dispersions. Only a few examples will be offered here.

a) Electrokinetics

One of the advantages of the new systems described in this work over the traditional

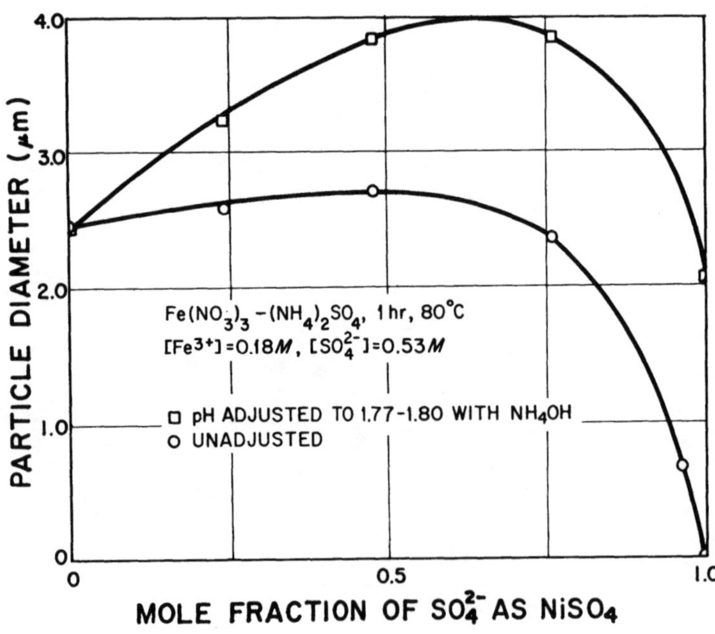

Fig. 14. Diameter of basic iron(III) sulfate particles as a function of mole fraction of $NiSO_4$ at pH adjusted to 1.77—1.80 ($\square$) or unadjusted pH ($\bigcirc$). The sols were prepared by aging solutions 0.18 M in $Fe(NO_3)_3$ and 0.53 M in total SO_4^{2-} concentration at 80 °C for 1 hour. The balance of sulfate was added as $(NH_4)_2SO_4$

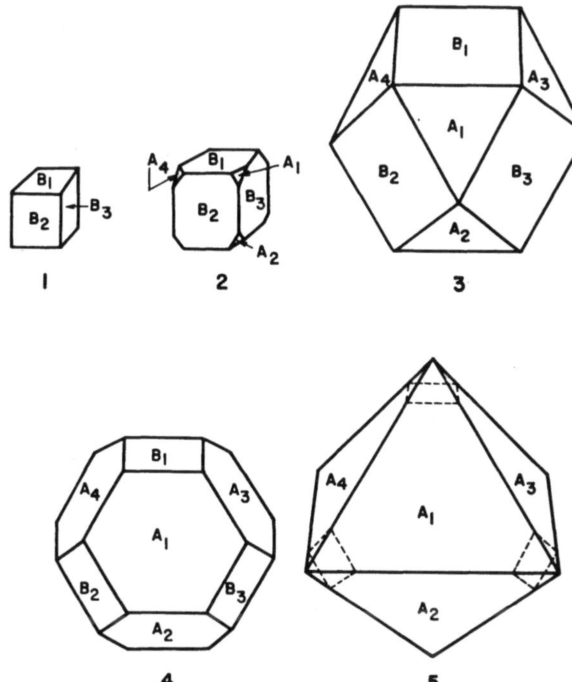

Fig. 15. Schematic diagram of various particles shapes of cubic symmetry.

latices is the ease with which the particle charge can be changed both in magnitude and in sign by simply varying the pH. This also infers that all of the metal hydrous oxide sols exhibit a zero point of charge (z.p.c.). The latter quantity is of great theoretical importance, yet the values reported in the literature for apparently the same metal hydrous oxides can vary appreciably (9). Electrokinetic measurements on chemically well defined monodispersed sols should yield reliable data on the z.p.c. and help resolve differences in reported values. Fig. 18 gives several electrokinetic mobility curves as a function of pH of colloidal dispersions prepared as given in the legend. One can see that the z.p.c. depends on the sol and that it varies considerably with the metal hydroxide studied. Two curves are shown for aluminum hydroxide; one is for the spherical particles containing sulfate ions, and the other for the same particles after the sulfate was exchanged for hydroxyl ions. A significant shift to the lower z.p.c. value is observed when the particles are sulfated. Interestingly, the z.p.c. values of 9.3 and 7.2 for the two systems studied represent essentially the upper and the lower limit of a large number of

data reported in the literature for aluminum hydroxide. This result clearly shows the considerable effect certain anions have on the charge of metal hydrous oxides.

b) Mechanism of formation of the chromium hydroxide sol

Hardly any documented quantitative information is available on the chemical processes involved in the precipitation of metal hydroxides. Recent work carried out with monodispersed chromium hydroxide (4) represents an effort in this respect, and it has produced evidence on the role of the solute complexes in the solid phase formation. Various species formed on aging of chromium salt solutions in the presence of sulfate ions were separated by paper electrophoresis. They were further characterized by means of radioactive tracers, primarily ^{51}Cr and

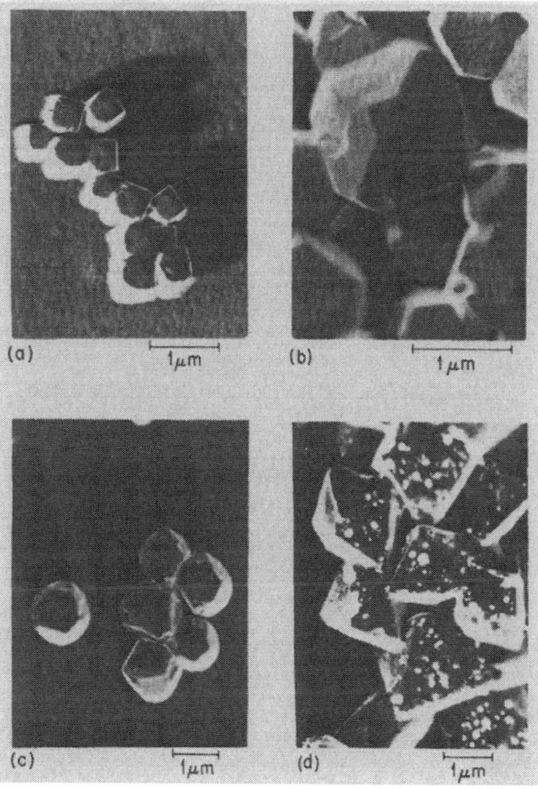

Fig. 16. Electron micrographs of replicas of copper(I) oxide sol particles showing the particle shapes, prepared from solutions of copper(II) salts of different concentrations (a) cubes, 5.5×10^{-4} M Cu^{2+}; (b) and (c) cuboctahedra, 2.8×10^{-3} M; (d) octahedra, 6.9×10^{-3} M. Glucose was present in all solutions

Fig. 17. Scanning electron micrograph of titanium dioxide sol particles prepared by aging for 41 days at $98\,^\circ$C a solution 0.106 M in $TiCl_4$ with a $[SO_4^{2-}]/[Ti^{4+}] = 3.8$

^{35}S. Fig. 19 shows that four kinds of species, according to their ionic charges, can be identified in an aged solution of chrome alum. Such radiopaper eletrophoretic chromatograms were obtained at different times of aging and for different electrolyte solutions. A quantitative analysis of the change in the concentration of various species has indicated that complexes "D" of zero charge and complexes "B" of $2+$ charge play the dominant role in the chromium hydroxide formation. All of the information points to the mechanism by which a complex chromium basic sulfate precursor forms first upon which the neutral solute hydrolysis product, $Cr(OH)_3$, nucleates and grows in a sequence of reactions

which may be represented by the following scheme:

$$m\,[Cr_2(OH)_2SO_4]^{+2} + m\,SO_4^{-2} \text{ and } n\,Cr(OH)_3 \cdot aq$$

$$[Cr(OH)SO_4]_m + n\,Cr(OH)_3 \cdot aq$$

$$[Cr(OH)_3]_{m+n}\,(\text{solid}) + m\,SO_4^{-2} + 2m\,H^+$$

This mechanism explains the role of sulfate ions and also the finding that the finally developed spherical particles of chromium hydroxide contain no sulfate. For a detailed account of the results which lead to the described mechanism the reader is referred to a recently published paper (4).

Monodispersed chromium hydroxide sols have also been utilized to determine the mechanism of the particle growth (10). For this purpose Nielsen's chronomal analysis was applied (11), which indicated that the growth occurs via surface reaction. The corresponding rate constants and energies of activation could also be evaluated.

c) Effects of polymers on particle growth

Numerous studies on the effects of surfactants on particle formation and growth have been reported with generally accepted conclusions that, when solids precipitate in the presence of surface active agents, the particle number

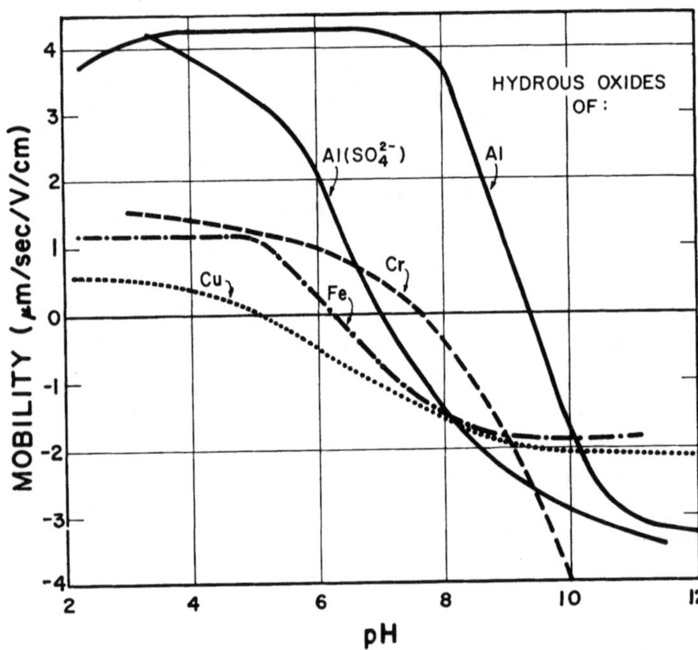

Fig. 18. Electrophoretic mobilities of different monodispersed metal hydrous oxide sols as a function of pH. Copper hydrous oxide (.....), basic ferric sulfate ($-\cdot-\cdot-$) chromium hydrous oxide ($----$), and aluminium hydrous oxide ($———$). The two curves for the last sol refer to original spherical particles containing sulfate ions, $[Al(SO_4^{2-})]$, and to spherical particles in which the sulfate ions were exchanged for hydroxyl group, $[Al]$

Fig. 19. Typical results of radio-paper electrophoresis on the electrolyte medium. A labelled solution initially 4.0×10^{-4} M in chrome alum was heated at 75 °C for 9 days. After cooling and filtering off the particles, the filtrate was subjected to paper electrophoresis

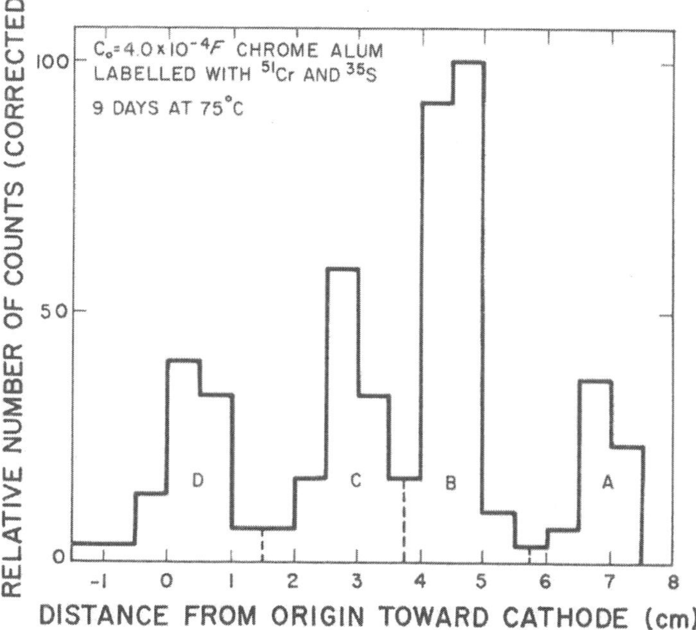

concentration is higher than in the absence of these additives; consequently, the average particle size is smaller. Much less work has been reported on the influence of polymers on particle formation and growth and the results are by no means conclusive (12).

The availability of monodispersed inorganic latices and the ease with which they can be prepared makes it now possible to systematically investigate the influence of polymers on particle formation and growth. Such a study has been undertaken with chromium hydroxide sols in the presence of various dextrans. These polymers are particularly convenient because they can be obtained in rather narrow molecule weight fractions in nonionic, anionic, and cationic modifications.

One example of this type of investigation is given in fig. 20, which shows the effect of dextran sulfate 2000 (molecular weight $\sim 4 \times 10^6$) on the particle size of chromium hydroxide. The four electron micrographs are for sols obtained in the absence and in the presence of three different concentrations of the polyelectrolyte. Obviously, an addition of only 0.001% (by wt.) of dextran sulfate 2000 has a considerable effect on the particle size; as the concentration of the polymer increases, the average particle size becomes smaller. Similar observation was made with the cationic dextran (DEAE) of the same chain length; however, the concentrations of this

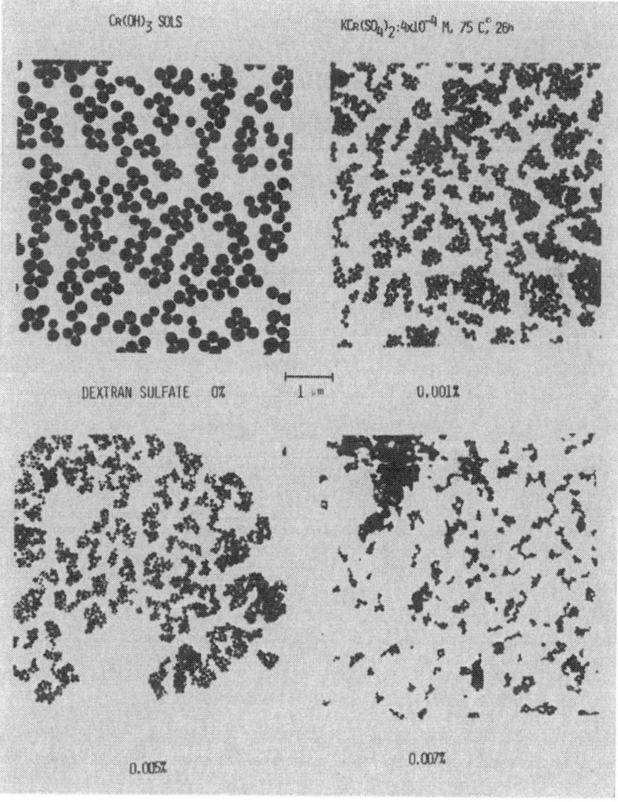

Fig. 20. The effect of a dextran sulfate 2000 on the particle size of a chromium hydrous oxide sol prepared by aging a 4×10^{-4} M solution of chrome alum at 75 °C for 26 hours. The four electron micrographs are for sol particles in the absence and in the presence of 0.001, 0.005, and 0.007% (by wt.) dextran sulfate, respectively

polyelectrolyte necessary to cause the same reduction of particle size were nearly two orders of magnitude higher than with dextran sulfate. Except in very high concentrations nonionic dextrans showed little influence on the formation of chromium hydroxide sols.

It was assumed that the rather significant effect of dextran sulfate on the particle size was due to ion binding of chromium ions by the polyelectrolyte. Dialysis equilibration studies on the same systems indeed showed a considerable depletion of chromium solute species on addition of the anionic polyelectrolyte. This change in solution concentration, thus, explains the difference in the size of the particles, since it was shown (fig. 1) that the dispersity of the sol depends on the concentration of the metal salt in solution.

The effect of the cationic polyelectrolyte, which is considerably smaller, is not understood at this time.

d) Heterocoagulation

Another application of the monodispersed inorganic latices has been found in the study of heterocoagulation. According to the current theories of coagulation of unlike sols, two parameters are of primary importance: the ratio of sizes and the ratio of surface potentials of the interacting particles. While the effect of varying sizes is easily explored with conventional organic latices, the variation of surface potentials are more difficult to accomplish. The latter problem can be overcome if the system used to study heterocoagulation consists of an organic and an inorganic latex. Indeed, an investigation has been conducted on a number of mixed sols consisting of monodispersed polyvinyl chloride (PVC) latices stabilized by fixed sulfate groups and monodispersed chromium hydroxide sols. The fact that both systems can be obtained in a variety of modal diameters, allows for the elucidation of the particle size effect. Furthermore, a change in pH has relatively little influence on the charge of the PVC latex; its sign remains the same and the value changes slightly, particularly over the higher pH range. On the other hand, the charge of the chromium hydroxide particles goes from positive to negative as the pH rises and the magnitude of the charge itself varies considerably. Thus, one can investigate the effects of surface potentials of unlike

particles over a wide range of conditions. The results on the kinetics of heterocoagulation of the described systems have been compared with the existing theories and found in excellent qualitative agreement, although quantitative correlation is still lacking (13).

By using different inorganic latexes, such type of studies can now be extended to include the effects of particle shape and other parameters of interest in the behavior of mixed colloidal dispersions.

Concluding remarks

The feasibility of the preparation of monodispersed metal hydrous oxide sols has now been amply demonstrated. It is recognized, that this is only the first step in the elucidation of the processes involved in the formation of metal hydroxides and in the characterization of their various properties. There is obviously one common underlying principle involved in the precipitation of metal hydroxides: metal ion hydrolysis. Since the latter process varies with each metal, one would expect the conditions for the precipitation of the corresponding hydroxides to be different. The fact that ions other than hydroxyl may coordinate with the hydrolyzable metal species further complicates the picture. Thus, we would expect different specific chemical reactions to take place in solutions from which the metal hydroxides precipitate. Indeed, the experience so far has shown that the monodispersed systems vary considerably in the chemical composition of the final products, as well as in the particle size, shape, charge, and crystallinity. One of the intriguing questions is why certain systems appear in the form of amorphous particles, whereas some others give well defined crystalline materials. As a first approximation, it would appear that metal ions known for their tendency to continue to polymerize on heating or aging (such as aluminum and chromium) finally give amorphous particles, whereas metal ions which hydrolyze to well defined mono- and polynuclear species yield crystalline solids. Anions which promote polymerization of metal hydroxylated species (e.g., sulfate) would be expected to cause amorphous particles to form, whereas anions which give well defined complexes (e.g., chloride) should produce crystalline solids. This has, indeed, been found to be the case with aged aluminum salt solutions.

It is our belief that the new inorganic latices described in this work will be employed in many fundamental investigations and that they will find numerous useful application. For the latter purpose, procedures will have to be developed for their preparation economically in large quantities.

Acknowledgement

This presentation has been based on the work of a number of my associates to whom I am indebted for their contributions. Specifically, I would like to express my appreciation to Dr. *A. Bell*, Dr. *A. Bleier*, Mr. *R. Brace*, Mr. *M. Budnik*, Mr. *R. T. Cataldi*, Dr. *D. L. Catone*, Dr. *R. Demchak*, Mr. *M. Dennen*, Dr. *S. Kratohvil*, Dr. *A. D. Lindsay*, Dr. *P. McFadyen*, Dr. *J. B. Melville*, Mr. *M. Onofusa*, Mr. *R. Sapieszko*, Mr. *P. Scheiner*, and Mr. *W. Scott*.

Summary

The methods for preparation of colloidal dispersions of metal hydrous oxides consisting of particles exceedingly uniform in size and shape are described in detail. These sols are, as a rule, prepared by aging acidic solutions of metal salts at elevated temperatures. It is shown that the anions play a decisive role in the particle formation. Electron micrographs of such systems, as obtained from solutions of aluminum, chromium, thorium, titanium, iron, and copper salts, will illustrate that certain metal hydrous oxides appear in the form of spherical (amorphous) particles while others give well defined crystalline particles of different shapes. Certain properties and applications of these new inorganic latices are illustrated and discussed.

References

1) *Demchak, R.* and *E. Matijević*, J. Colloid Interface Sci. **31**, 257 (1969).
2) *Matijević, E., A. Bell, R. Brace*, and *P. McFadyen*, J. Electrochem. Soc. **120**, 893 (1973).
3) *Matijević, E., A. D. Lindsay, S. Kratohvil, M. E. Jones, R. I. Larson*, and *N. W. Cayey*, J. Colloid Interface Sci. **36**, 273 (1971).
4) *Bell, A.* and *E. Matijević*, J. Inorg. Nucl. Chem. **37**, 907 (1975).
5) *Brace, R.* and *E. Matijević*, J. Inorg. Nucl. Chem. **35**, 3691 (1973).
6) *Matijević, E., R. S. Sapieszko*, and *J. B. Melville*, J. Colloid Interface Sci. **50**, 567 (1975).
7) *McFadyen, P.* and *E. Matijević*, J. Colloid Interface Sci. **44**, 95 (1973).
8) *Budnik, M.*, M. S. Thesis, Clarkson College of Technology (1975).
9) *Parks, G. A.*, Chem Rev. **65**, 177 (1965).
10) *Bell, A.* and *E. Matijević*, J. Phys. Chem. **78**, 2621 (1974).
11) *Nielsen, A. E.*, Kinetics of Precipitation (Oxford 1964).
12) *Lindsay, A. D., E. Matijević*, and *J. P. Kratohvil*, Colloid & Polymer Sci. **253**, 581 (1975).
13) *Bleier, A.* and *E. Matijević*, J. Colloid Interface Sci. (June 1976).

Author's address:

Egon Matijević
Institute of Colloid and Surface Science
and Department of Chemistry
Clarkson College of Technology
Potsdam, New York 13676, USA

Progr. Colloid & Polymer Sci. **61**, 36—45 (1976)
© 1976 by Dr. Dietrich Steinkopff Verlag GmbH & Co. KG, Darmstadt
ISSN 0340-255 X

Plenary lecture of the IUPAC-Conference on Colloid and Surface Science in
Budapest, September 15—20, 1975

*„Ruder Bošković" Institute, Zagreb, and Laboratory of Physical Chemistry, Faculty of Sciences,
University of Zagreb, Zagreb (Yugoslavia)*

Adsorption of iodide ions and nucleation of freshly prepared silver iodide sols *)

M. Mirnik and *S. Musić***)

With 7 figures

(Received December 9, 1975)

Introduction

The measurements of the adsorbed constituent I^- ions in various forms of AgI precipitates represent important quantitative information on the double layer of "ionic solid-electrolyte solution" systems of colloid chemistry and interface science. The adsorbed amount of I^- ions on AgI can be conveniently determined by the potentiometric method. This method was used several times in the determination of the surface charge of preformed dried AgI precipitates (1). The adsorbed I^- ions to the best knowledge of the authors have not been investigated in recent years by the same method on freshly precipitated forms of AgI except in ref. (2). The same publication gives the hystorical background of the development of the potentiometric method and reports on the results obtained which can be summarized as follows:

1. The adsorbed amount of I^- ions per mol of fresh stable AgI, γ_{stab}, decreases linearly with pI. The slope represents the molar adsorption capacity Γ_{stab} and can be expressed conveniently by $\Gamma = 100\, \gamma_{stab}/[AgI]^* \Delta pI$ in mol%, where $[AgI]^*$ is the amount of AgI in mol dm^{-3} of sol. It can be considered the sol concentration. The negative logarithm of the I^- activity is pI, its

change ΔpI and γ_{stab} is the amount of adsorbed I^- ions per mol of stable AgI.

2. On fresh stable AgI prepared in a sol concentration *) $[AgI] = 1.0 \times 10^{-3}$ mol dm^{-3} the slope is higher in the range of $5 < pI < 8$ than in the range of $8 > pAg > 5$, the change of the slope being abrupt at $pI = pAg = 8$.

3. The adsorption capacity in mol% decreases from about 0.3 to 0.1 when the concentration increases from $[AgI]^* = 1.0 \times 10^{-4}$ to 5.0×10^{-2} in the region of $5 < pI < 8$ and from 0.12 to 0.06 in the region of $8 > pAg > 5$.

4. The increase of the concentration from $[NaNO_3]^* = 1.0 \times 10^{-3}$ to 1.0×10^{-2} causes a relatively small decrease of the adsorption capacity from 0.27 to approximately 0.16, while in passing the coagulation value of the Na^+ ion, i.e. from $[NaNO_3]^* = 0.09$ to 0.1 the capacity decreases significantly to about 0.06 mol%.

5. The linear decrease of the adsorbed I^- is accompanied by an increase in the size of the stable particles from ~ 10 nm to ~ 150 nm observed by light scattering, obviously indicating their clustering.

The aim of the present investigation was to obtain further quantitative information on the influence of the AgI sol concentration, its aging, and coagulation by counter ions of different valencies, on the quantity of adsorbed I^- ions presented as function of pI.

The potentiometric method used in the determination of the adsorbed constituent I^- ions

---

*) Contribution No. 169 of the Laboratory of Physical Chemistry.

**) Results taken in part from the M. Sc. thesis by *S. Musić*.

was the same as in ref. (1). It gave the "adsorbed I⁻ ion-pI" plot as the result of the differences in the titration curves in the presence and absence of AgI.

Experimental

Apparatus. A potentiometer/Derriton Instruments Ltd. England, Catalog N. E 4248/ coupled with a galvanometer as zero instrument (7000 ohm resistance, sensitivity 6×10^{-10} A/mm) was used for the measurement of the electromotive force (EMF) of the cell. The Ag, Ag_2S indicator electrode (Ag wire, 0.05 cm diameter, 2.5 cm exposed length, melted in a glass capillary) was prepared by anodic polarization with a current of 0.5 mA during 10 minutes of slow rotation. The cathode was a Pt wire, the electrolyte a $[Na_2S]^* = 10^{-2}$ solution. The reference electrode was a Ag/AgBr electrode, anodically prepared in a $[KBr]^* = 10^{-2}$ solution. The sols and the reference electrode solution were separated by a $[KNO_3]^* = 10^{-3}$ solution which was in a glass attachment with an asbestos filament melted in it (to ensure the contact of electrolytes without mutual contamination).

Thus, the cell was set up according to the scheme:

Ag, AgBr | $[KBr]^* = 10^{-2}$ | $[KNO_3]^*$
$= 10^{-3}$ | AgI, I⁻ sol | Ag_2S, Ag

The Ag, Ag_2S electrode was used instead of the paraffined Ag, AgI electrode which was used in ref. (1). In this way some of the results of ref. (1) were reproduced by an independent technique. The Ag, Ag_2S electrode proved to be reproducible, easy to prepare and reliable for longer periods of time. Surfactants and various electrolytes did not significantly affect its response to I⁻ and Ag⁺ ions in the range of pI > 5 to pAg > 5.

Reagents. The stock solutions of $AgNO_3$, KNO_3, $Ba[NO_3]_2$, $La[NO_3]_2$ and KBr were prepared by dissolution of weighed quantities of dried salts. The exact concentration of the NaI solution was determined by titration with the exact $[AgNO_3]^* = 1.000 \times 10^{-3}$ standard solution. The solutions of lower concentrations were obtained by dilution of 0.1 mol dm⁻³ stock solutions. Chemicals of the highest analytical purity and twice distilled water were used for dissolution and dilution.

Preparation of sols and suspensions. To a predetermined volume of water, NaI and neutral electrolyte solution, agitated with a magnetic stirrer an $AgNO_3$ solution was added from a graduated pipette at a constant rate of about 10^{-4} mol dm⁻³ min⁻¹ till a sol of a given concentration and pI = 5 was obtained. The adsorption was measured in the volume of 100 cm³ of fresh sols 5 minutes after pI = 5 was obtained in the sols. The adsorption of aged sols was measured at given

*) In the present paper all concentrations are given in mol dm⁻³ and are designated conveniently by the compound formula in square brackets with an asterix e.g. $[AgI]^* = 1 \times 10^{-3}$, $[NaNO_3]^* = 1 \times 10^{-2}$ instead by $[AgI] = 1 \times 10^{-3}$ mol dm⁻³, $[NaNO_3] = 1 \times 10^{-2}$ mol dm⁻³.

times on 100 cm³ samples of sols originally prepared in a volume of 1 dm³.

Potentiometric titration. The variation of the adsorbed amount of I⁻ with pI-pAg, the concentrations of I⁻ or Ag⁺ ions and the theoretical behaviour of electrodes were determined by potentiometric titrations which were performed in the following way: To 100 cm³ of the test solution in a 150 cm³ beaker $[AgNO_3]^* = 1.00 \times 10^{-3}$ solution was added from a horizontal buret of 1 cm³ volume. The indicator and the reference electrode were dipped in the test solution and the EMF of the cell was measured 3 minutes after each addition of $AgNO_3$ solution. During this time the readings became constant except in the pI = 7 to pAg = 7 range. In this range the readings were taken after several minutes when the shift of the EMF became less than $1-2$ mV per minute.

Check of the electrodes. The electrodes were checked by titration of a

$[NaI]^* = 1.00 \times 10^{-5} - [NaNO_3]^* = 1.00 \times 10^{-3}$

solution with a $[AgNO_3]^* = 1.000 \times 10^{-3}$ solution. The electrodes were considered satisfactory when linear plots could be drawn through the points at pI = 5 and pAg = 5 with a slope of approximately $\Delta E/\Delta pI = 59 \pm 6$ mV in the "EMF-pI/pAg" diagram. As a rule a correction of the analytical equivalent volume of not more than $\pm 0.2\%$ was made for each titration in order to obtain linearity in the range pI = 7 to pAg = 7.

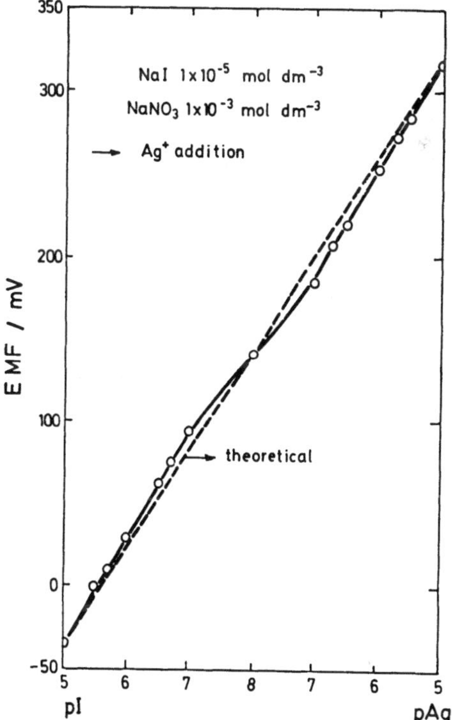

Fig. 1. The dependence of the potential of the Ag/Ag_2S electrode on the activity of I⁻ and Ag⁺ measured as the EMF of the cell:

Ag, AgBr | $[KBr]^* = 10^{-2}$ | $[KNO_3]^*$
$= 10^{-3}$ | I⁻ (Ag⁺) | Ag_2S, Ag
asbestos

Such corrections are justified by the error of the volume determination in each titration. A typical check of the electrode given by the linear "pI-pAg against EMF" plot is given in fig. 1. The lines on the pI and pAg side are shifted by a certain value because the solubility product is not exactly 10^{-16} as assumed when the pI and pAg scales meet at the value 8. The diffusion potential was not completely eliminated either, and it can influence the slope and cause a shift of the lines on both sides of the diagram.

Results

Plot of "I⁻ adsorbed against pI-pAg". The variation of the I⁻ adsorbed, i.e. of γ with pI was derived from the titration plots of the negative sols i.e. from the plots "EMF against Ag⁺ added". In fig. 2 four plots are given for sol concentrations $[AgI]^* = 1 \times 10^{-3}$, 2.5×10^{-3}, 5.0×10^{-3} and 10.0×10^{-3}. The excess concentrations of I⁻ and Ag⁺ on the abscissa were calculated from the amount of $AgNO_3$ added assuming that at pAg $= 5.45$ the adsorbed amount equals zero. Linear portions of the plots are evident, especially at higher concentrations of AgI. The amount at a given EMF adsorbed was obtained from the same plots by subtraction of the excess of I⁻ or Ag⁺ which was necessary to produce the same EMF, with no AgI present. The "I⁻ adsorbed against pI" plot is therefore the difference between the titration curves of the sol and of the $[NaI]^* = 10^{-5}$ solution. The theoretical titration plot "EMF logarithm excess concentration of I⁻ (or Ag⁺)" for each electrode had a slope of $\Delta EMF/\Delta pI = 59 \pm 6$ mV

and passed the calibration points obtained with pI $= 5.0$ and pAg $= 5.0$ solutions (fig. 1). The calibration values were, when necessary, corrected for ± 2 mV to give straight lines as far as possible towards low pI values. This correction is justified by the deviation from the measured potential of the actual calibration value in a given titration. The absolute zero point of adsorption could not be determined from titrations; therefore it was assumed to be at pAg $= 5.45$. The amount adsorbed per liter divided by the final AgI concentration gives the adsorbed amount in mol I⁻ per mol AgI. The results are presented in fig. 3. Thus, the plots of fig. 3 were shifted vertically in such a way that the zero adsorbed amount, i.e. the zero point of adsorption, was at pAg $= 5.45$. The abscissa values are the potentials measured while the pI-pAg values were assigned on the basis of the calibration points in the pI $= 5$ and pAg $= 5$ points. The distance between these two points was divided in 6 equal parts to give the points at pI $= 6, 7, 8$ and pAg $= 8, 7$ and 6. The points pI $= 8$ and pAg $= 8$ coincide, which means that the assumption was made that the solubility product of AgI is 10^{-16} exactly.

Fig. 3 demonstrates the influence of the sol concentrations $[AgI]^* = 1.0 \times 10^{-3}$, 2.5×10^{-3}, 5.0×10^{-3} and 1.0×10^{-2} upon the "I⁻ adsorbed against pI-pAg plot" in mol I⁻ per mol AgI in the presence of $[NaNO_3]^* = 1.0 \times 10^{-2}$.

The adsorbed amount mol I^-_{ads} per mol AgI is the highest for the sol precipitated in the lowest investigated concentration $[AgI]^* =$

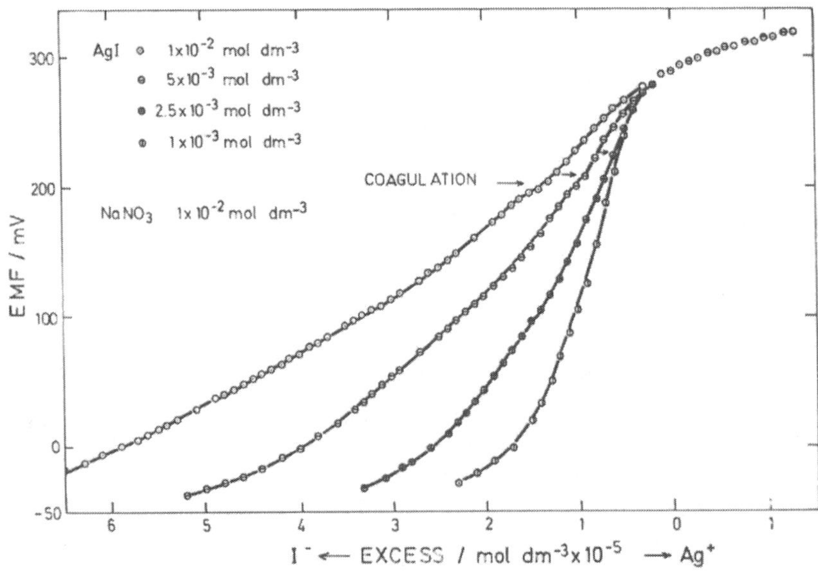

Fig. 2. The influence of the concentration of AgI sols, freshly precipitated ("in statu nascendi") upon the potentiometric titration plots given as "excess I⁻ or Ag⁺ against EMF". Sol concentration $[AgI]/mol\ dm^{-3} = 1 \times 10^{-3}, 2.5 \times 10^{-3}, 5 \times 10^{-3}, 1 \times 10^{-2}$, $[NaNO_3]/mol\ dm^{-3} = 1 \times 10^{-2}$

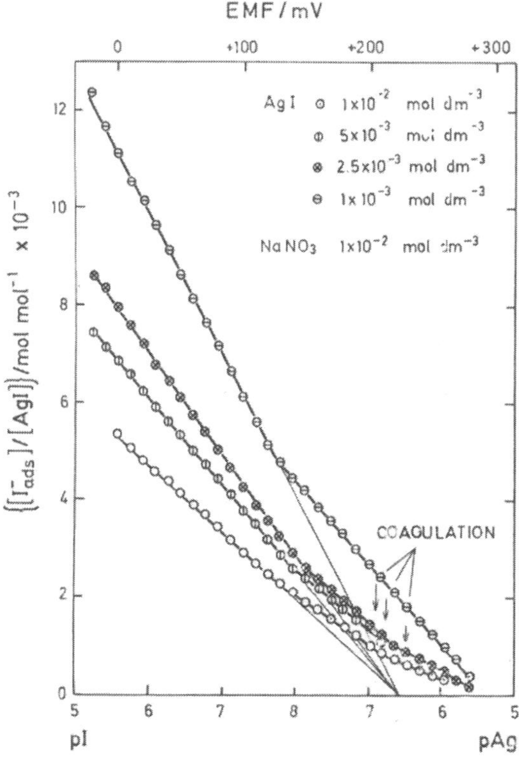

Fig. 3. The influence of the concentration of freshly precipitated AgI sols upon the plots of "adsorbed I⁻ mol per mol AgI × 10⁻³ against pI-pAg (EMF)". Sol concentrations as in fig. 2

1.0×10^{-3} at which concentration the plots have two linear parts. One line is in the range of $5 < \mathrm{pI} < 8$ and the other with a smaller slope in the range of $8 > \mathrm{pAg} > 5$. There is a marked discontinuity in the slope at $\mathrm{pI} = \mathrm{pAg} = 8$. In the remaining three higher concentrations of AgI the amount I⁻ adsorbed is the lower the higher is the sol concentration. An additional discontinuity to that at $\mathrm{pI} = 8$ can be observed at approximately $\mathrm{pAg} = 6.6$ to 6.9. Approximately, $\mathrm{pAg} = 6.6$ to 6.9 represent the values at which the sol suddenly flocculates after the last portion of added Ag⁺ to form visible flocks. This point, precisely determined by light scattering observation was called the "negative activity limit of stability". The choice of the real zero point of adsorption at pAg 5.45 caused a single intersection at zero adsorbed amount and at pAg 6.6 also of the lines measured in the range $5 < \mathrm{pI} < 8$ extrapolated to zero. The variation of the amount adsorbed of I⁻ per mol AgI can therefore be expressed by the formula

$$\gamma = \Gamma(\mathrm{pI}^0 - \mathrm{pI})/100 \qquad [1]$$

where Γ is the adsorption capacity, i.e. the slope of the same plots calculated by the formula

$$\Gamma = \Delta\gamma \, 100/\Delta\mathrm{pI} . \qquad [2]$$

It is certain that the above formula is valid also in the range below $\mathrm{pI} = 5$. $\mathrm{pI}^{0\prime} = 9.4$ is the extrapolated zero point of adsorption of the lines measured in the range of $5 < \mathrm{pI} < 8$, and $\mathrm{pI}^{0\prime\prime} = 16 - 5.45 = 10.55$ is the assumed zero point of adsorption for the range $\mathrm{pI} > 8$. The fact that the extrapolated lines have a single intersection, i.e. a single zero point confirms that the assumed value of the zero point at $\mathrm{pI}^{0\prime\prime} = 10.55$ is correct in principle.

The influence of aging at $\mathrm{pI} = 5$ and the sol concentration upon the adsorption capacity in the range of $5 < \mathrm{pI} < 8$ is demonstrated by fig. 4. The adsorption capacity per mol AgI is the higher the lower is the sol concentration and the fresher the sol. It follows from fig. 4 that a sol freshly precipitated in a high concentration (10^{-2}) has approximately the same adsorption capacity $(\Gamma \sim 0.09 - 0.14 \,\mathrm{mol\%})$ as an aged sol

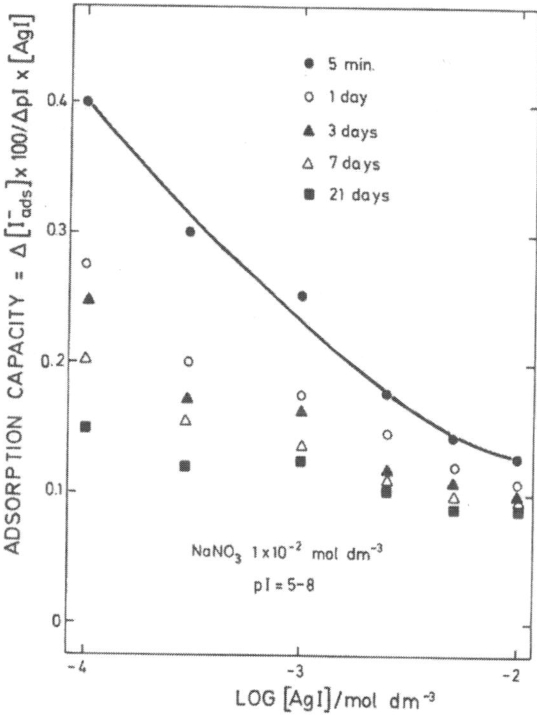

Fig. 4. The dependence of the molar adsorption capacity upon the sol concentration and age (= 5 minutes, 1, 3, 7 and 21 days)

precipitated in a low concentration. Also, the relative influence of the aging is much more pronounced in low sol concentrations than in the high ones. The adsorption capacity of approximately 0.1 mol% seems to be the limiting value produced by sufficient aging or by sufficiently high sol concentrations. The adsorbed amount of sols prepared in concentrations $[AgI]^* = 10^{-3}$ to 10^{-2} and aged for over 3 weeks can be approximately calculated by the formula

$$\gamma_{\text{stab, aged}} = 0.1 \, (9.4 - pI)/100 . \qquad [3]$$

The influence of the increasing concentrations of the Ba^{2+} ion on the "Ag^+ added against EMF" plot and "I^- adsorbed against pI" plot is demonstrated in fig. 5. The later plots are linear. The Ba^{2+} ion is used as an example and analogous plots were obtained with La^{3+} and Na^+. The higher the concentration of the counter ion, the lower is the slope of the plot. Consequently, the adsorbed amount of I^- at a given pI is smaller. The lowest slopes were obtained with cation concentrations higher than their coagulation values $([1/2 \; Ba^{++}]^*_{\text{coag}} = 3.8 \times 10^{-3})$. It is most

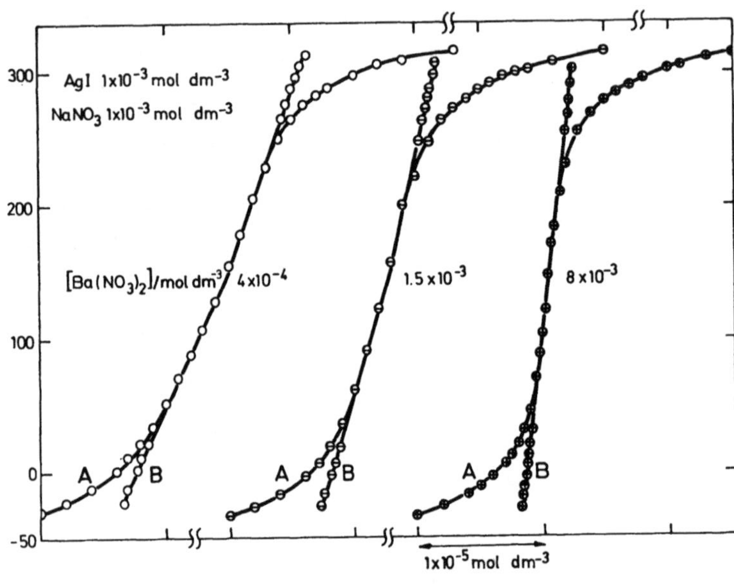

Fig. 5. Influence of Ba^{2+} ion concentration upon the plots of "excess I^- (Ag^+) against EMF" and "adsorbed amount of I^- against EMF". $[AgI]/mol \; dm^{-3} = 1 \times 10^{-3}$, $[Ba(NO_3)_2]/mol \; dm^{-3} = 4 \times 10^{-4}$, 1.5×10^{-3}, 8×10^{-3}

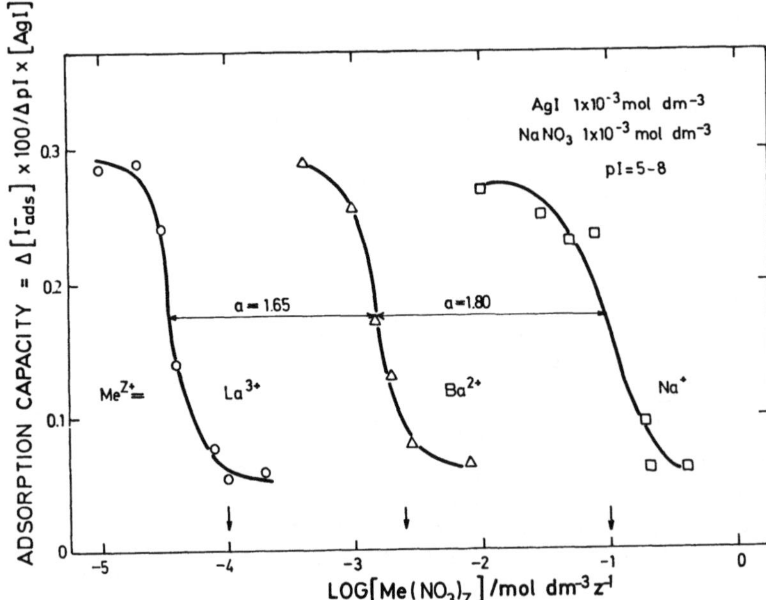

Fig. 6. Influence of the $1/3 \; La^{3+}$, $1/2 \; Ba^{2+}$, $1/1 \; Na^+$ concentration upon the adsorption capacity in the range $5 < pI < 8$. $[AgI]/mol \; dm^{-3} = 1 \times 10^{-3}$, $[NaNO_3]/mol \; dm^{-3} = 1 \times 10^{-3}$

probable that the counter ions influence, besides the slope, also the zero point of adsorption. However, with the present technique it was not possible to obtain the zero point of adsorption for coagulated suspensions either. The usual precision of the determination of the absolute amounts of Ag^+ and I^- present in the system, does not permit to establish the equivalency between the two ions with a high enough precision. At the lower sol concentrations of $[AgI]^* = 1.0 \times 10^{-3}$ the difference between the slopes in the $pI < 8$ and $pI > 8$ ranges is too small to be observed.

The slopes of the plots in fig. 5 in mol% i.e. the adsorption capacities are given in fig. 6 as functions of the logarithm counter ion concentrations for the counter ions Na^+, Ba^{2+}, La^{3+} each representing its valency group as example. In the narrow logarithm concentration range of less than 1, in the vicinity of the coagulation values which are shown by vertical arrows, the capacity decreases from the high value of 0.3, to the low values of coagulated suspensions of ~ 0.06 mol%.

Discussion

The present investigation of I^- adsorption on AgI represents one of the few known in which the AgI sols were freshly prepared and consequently of the smallest possible particle size. Although it is known that the freshly prepared sols are not monodisperse (4), there is no doubt that the range of the particle sizes is usually within given limits and one or two average values of the size and perhaps an average geometrical form can be assumed for the sols which are prepared in a controlled way. To check any double layer theory, data on charge density, i.e. on the number of ions adsorbed per unit surface area or per particle are required. However, there are no sufficiently precise methods known by which the size or the surface of the particles and the particle size distribution could be determined for fresh polydisperse AgI sols. The fresh sol particles are too small to be determined by visible light scattering techniques (1—50 nm) in situ. Dye adsorption methods cause changes in dispersity and coagulation. Techniques based on evaporation of samples or separation of the solid from liquid also include changes in dispersity. In addition to that, all fresh sols are mixtures of the hexagonal and cubic modification (3), and at the same time mixtures of two

forms of AgI having a different response in heterogeneous exchange of I^- ions caused by counter ions in coagulating concentrations (5). That is: one of the forms exchanges all its constituent ions with those in the liquid within a few minutes or less, while the other form does not exchange for longer periods of time, days or weeks, or does not exchange at all. All this means that in freshly prepared sols AgI is present in the form of particles of different sizes, different crystallographic modification and/or exchange properties. In addition to that during the experiments, the particles change their size and occasionally coagulate owing to the variation of pI, age and electrolyte concentration.

For all these reasons the results on I^- adsorption reported in the present paper are given in "mol adsorbed per mol AgI" instead of in "mol I^- adsorbed per unit area" or "adsorbed I^- charge in coulombs per unit area" of the particles. The results given as mol I^- adsorbed per mol AgI are unambiguous and perfectly reliable because the amount of AgI is exactly known.

It follows that it is impossible at present to measure continuously and in situ the change of the size or surface of the particles as a function of pI. This represents a prerequisite for the correlation of the variation of the adsorbed amount with the variation of the surface or size of the particles during the change of pI.

However, despite all these complicating factors, which include polydispersity, two crystallographic modifications and a minimum of two AgI forms of different heterogeneous isotope exchange properties, a strictly reproducible linear decrease of the adsorbed amount of I^- per mole of AgI (and also per litre of sol) is observed in the present paper (fig. 3).

In the lower sol concentration range $[AgI]^* = 10^{-4}$ to 10^{-3} the plots have two different slopes (the plot for $[AgI]^* = 1 \times 10^{-3}$ is shown only) and in the higher range $[AgI]^* = 2 \times 10^{-3}$ to 5×10^{-2} three slopes. In sols of all concentrations the adsorbed amount per mol, γ, and the slopes Γ diminish with higher concentrations and at $pI = 8$ the slopes change abruptly.

Rapid coagulation does not cause any observable changes of the slope [2] in the low sol concentrations. Rapid coagulation can be observed through increase in the intensity of the scattered light or through the formation of visible flocks when pAg passes the $pAg_{limit} =$

6.2—6.5 value, i.e. the so-called negative stability limit value.

In the higher sol concentration range of $[AgI]^* = 2 \times 10^{-3}$ to 5×10^{-2} at pAg_{limit} a second decrease of the slope can be observed. After a given addition of Ag^+ ions to the sols, during the formation of visible flocks, pAg increases observably with time. This means that the adsorbed amount of I^-_{ods} decreases measurably in the first minutes after the pAg_{limit} value is passed.

During the linear decrease of the amount adsorbed, caused by each increase of pI, the particles increase, as it was proved in ref. (2) by the observed increase of the intensity of scattered light. This increase can be taken to explain the linear plots. One should conclude therefore that during the increase of pI, the average size of the particles increases in a very regular way.

All the systems in the range between $pI = 5$ and $pAg = 6.6$ (for $[NaNO_3]^* < 10^{-2}$) remain stable for months. The supernatant solution is always turbid and the greater part of the sol particles which sediment can be redispersed by repeated shaking. The plots in the range $6.6 > pAg > 4.5$ correspond to systems which are coagulated by the so-called isoelectric coagulation. The systems were left to stand for several hours (ref. (6)), the adsorbed amount of counter ions becomes equal to zero if counter ions are present in concentrations higher than their coagulation values. During the same time the pAg value shifts to higher values. This shift is obviously due to the decrease of Ag^+ concentration caused by the neutralization of adsorbed I^- ions with Ag^+ ions or is due to the release of I^- ions by aging i.e. by coarsening of the particles (ref. (2), fig. 1). The specific surface area of aged particle is smaller and therefore the adsorbed amount of I^- is also smaller.

The plots of fig. 3 are given in the range of $pI > 5$. In this range only the adsorbed amount of I^- can be determined, since the absolute concentration of I^- in the liquid is small in comparison with the absolute concentration of adsorbed I^-. It is most probably correct to assume, if measurement were possible that the linear plots would be valid, up to the highest possible concentrations (fig. 7). The highest possible concentration in a sol is that of the starting I^- ion concentration $[I^-]_s$. By the

addition of the $AgNO_3$ solution the volume of the sol significantly increased in some experiments and therefore the logarithm of the final sol concentration, $\log [AgI]_s$ differed occasionally from the logarithm of the starting I^- concentration $(-pI_s)$.

The value $pI_s = -\log[I^-]_s$ on the abscissa determines the maximal value γ_s on the extrapolated γ line (fig. 7). The crucial observation is that the γ_s values for each pI_s value are on a straight line with a defined slope Γ_s. The intersection of the latter line with abscissa is the zero point of adsorption pI_s^0, at which is $\gamma_s = 0$.

The linearity of the plot γ_s against pI_s is a definite indication that the parameter of the experiment is the starting pI_s value, and not only, as it was assumed during the experimentation, the final sol concentration $[AgI]_s$.

The interrelation between $[I^-]_s$ and $[AgI]_s$ is given by

$$[I^-]_s = [AgI]_s (v_{I^-,s} + v_{Ag^+})/v_{I^-,s} \qquad [4]$$

where $v_{I^-,s}$ is the volume of the starting I^- solution to which the volume v_{Ag^+} of the $AgNO_3$ solution was added.

The following considerations can be made.

The variation of the adsorbed amount can be represented by the linear equation

$$\gamma = \Gamma(pI^0 - pI) = \gamma_s - \Gamma(pI - pI_s). \qquad [5]$$

The slope of the line is defined by

$$\Gamma = \gamma_s/(pI^0 - pI_s) \qquad [6]$$

while the maximal amount adsorbed γ_s is defined by

$$\gamma_s = \Gamma_s(pI_s - pI_s^0). \qquad [7]$$

It follows that the adsorbed amount γ is determined by: (a) three constants, Γ_s, pI_s^0 and pI^0, which are characteristic of the system, (b) the parameter of the experiment represented by pI_s, and (c) the variable parameter pI.

Consequently, a theory to explain the above observations is proposed as follows.

The first addition of Ag^+ to a solution of $pI = pI_s$ however small it may be, yields practically a concentration of AgI which is much higher than the equilibrium saturation concentration for ionic and complex solubility of AgI. The saturation concentration is immediately surpassed perhaps by orders of magnitude. Therefore, immediately after the first addition

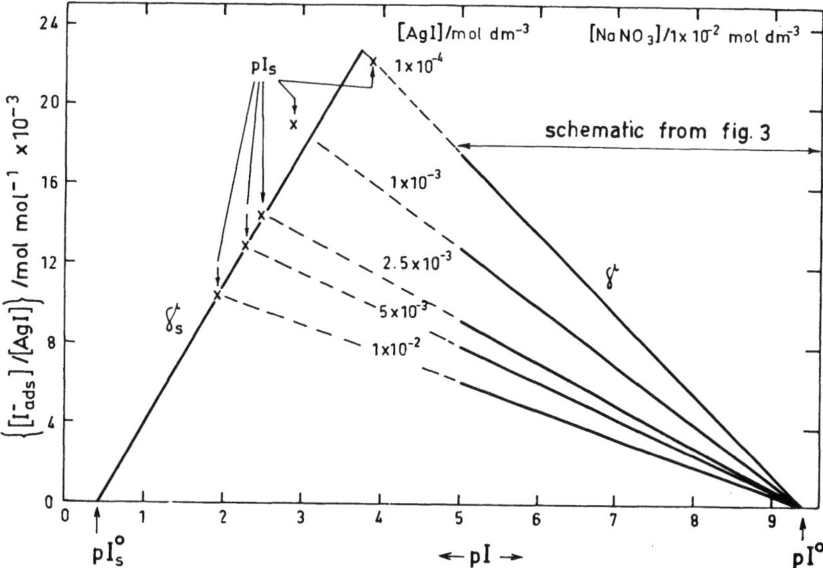

Fig. 7. The derivation of the γ_s plot from the measured γ plots and initial pI_s values

embryos are formed and their character is defined by the slope $\gamma_s/(pI_s^\circ - pI^0)$. Immediately after the first addition of Ag⁺ ions a given number j of embryos or nuclei is formed having the average number p of AgI units and ν adsorbed I⁻ ions. On further addition of Ag⁺ four processes take place. One is the growth of embria or nuclei to the final size of primary particles, i.e. the increase of $p \to p_s$, the second is the increase of the number of embria or nuclei $j \to j_s$, and the third the gradual neutralization of the adsorbed I⁻, i.e. the decrease of ν. The first two processes, i.e. the growth of the nuclei and the increase of their number represent in fact the precipitation of AgI which at the same time causes the increase of pI. The fourth is the formation of clusters of primary particles.

It was demonstrated by light scattering experiments, that the stable particles in the sols increase with pI. There is no doubt that the particles which cause the scattering of light are formed by aggregation of smaller particles and that in a sol there is a range of particles whose sizes are distributed statistically around one or two mean values. One can assume that each stable scattering particle is a cluster of a given number of primary particles and that the average number of primary particles in the clusters is proportional to the number of neutralized I_{ads}^- ions, i.e. to the value γ which reflects the value ν_{av}. This means that the third process caused by the addition of Ag⁺ to a sol is the neutralization of I_{ads}^- evidenced by the decrease of γ and that this process is accompanied by the

fourth process i.e. the aggregation of primary particles into stable clusters or aggregates of primary particles of an average size.

That the abrupt change of the slope at the $pI = 8$ (fig. 3) is caused by the abrupt decrease, per change in pI, of the number of primary particles which form clusters can be proved by an exact calculation based on the fact that at $pI = 8$ the number of particles with the average charge is equal for both slopes.

Also, it is possible to calculate the number, j_s, of primary particles in a sol of a given pI_s value and the maximal number of AgI units, p_s, in a primary particle because of the equations

$$j_s\, p_s = [AgI]_s\, L \qquad [8]$$

$$j_s\, \nu_s = \gamma_s [AgI]_s\, L \quad \text{and} \qquad [9]$$

$$\nu_s/\nu_{limit} = \gamma_s/\gamma_{limit} \qquad [10]$$

are valid. Here L is the Avogadro-Loschmidt constant.

If one assumes that the smallest possible charge per stable primary particle at pAg_{limit} is $\nu_{limit} = 1$ then $\nu_s = 10$ is the maximal charge in all three cases. Consequently,

$$p_s = \nu_{limit}/\gamma_{limit} = 1/\gamma_{limit} \quad \text{and} \qquad [11]$$

$$j_s = [AgI]_s\, \gamma_{limit}\, L . \qquad [12]$$

From fig. 3 and fig. 7 for the sols $[AgI]_s^* = 10^{-4}$, 10^{-3}, 10^{-2} and $\gamma_s = 22.2 \times 10^{-3}$, 18×10^{-3}, 10×10^{-3} $[I_{ads}^-]^*/[AgI]^*$ and $\gamma_{limit} = 2.2 \times 10^{-4}$, 1.8×10^{-3}, 1×10^{-3} $[I_{ads}^-]^*/[AgI]^*$ it follows that the numbers of the primary particles j_s are

1.3×10^{17}, 1.1×10^{18}, 6×10^{18} dm^{-3} and the numbers of AgI units p_s per particles are 450, 540, 1000 for the given sol concentrations.

Fig. 5 represents the transition of the adsorption plots from those characteristic of zero or low concentrations of a coagulating counter ion to those characteristic of high concentrations. The slope Γ can be used as a relative measure of the specific adsorbed amount which is approximately proportional to the slope because the shifts of pI^0 are small. A systematic representation of the influence of the counter ion concentration and valency is given in Fig. 6. There is a shift of the plots towards higher concentrations when the valency of the counter ion is lower. The shift of the plots is designated by $\underline{a}$ and is equal for the La^{3+}-Ba^{2+} plots

$$\underline{a} = \log \{[1/2 \, \text{Ba}^{2+}]/[1/3 \, \text{La}^{3+}]\}_{\Gamma = \text{const}} = 1.65 \quad [13]$$

and for the [Ba^{2+}-Na$^+$] plots

$$\underline{a} = \log \{[\text{Na}^+]/[1/2 \, \text{Ba}^{2+}]\}_{\Gamma = \text{const}} = 1.85 . \quad [14]$$

Practically the same value is obtained for the shift of the intersections of the tangents through the inflection points and the horizontal tangents through the minimum adsorption capacity.

Since a logarithm activity factor of $-\log f \doteq 0.15$ can be expected, in the high ionic strength of the [Na$^+$]$^* > 0.05$ concentrations one can conclude that the parameter $\underline{a}$ is practically equal for the valency combinations $1-2$ and $2-3$. For this reason one can express the rule for the influence of the valency of the counter ions by the expression

$$\log (z \, a_{M^z}/a_{M^0})_{\Gamma = \text{const}} = - z \, \underline{a} \quad [15]$$

where $(a_{M^z})_{\Gamma = \text{const}}$ is the transition activity (mol dm^{-3}) from the high Γ values of stable sols to the low values of coagulated suspensions, i.e. it is the counter ion activity at which

$$\Gamma = (\Gamma_{\text{high}} - \Gamma_{\text{low}})/2 = \text{const.} \, \log (a_{M^0})_{\Gamma = \text{const}}$$

is the constant of the linear equation, $\underline{a}$ is the proportionality constant, and z, the valency of the counter ion. The valency z is always taken positive for cations and anions. The analogous rule for the influence of the valency of the counter ions upon the coagulation values is called the Schulze-Hardy rule. Eq. [15] can therefore be considered to represent the linear formulation of the Schulze-Hardy rule for adsorption capacities. Since experimental values $\underline{a} = 1,65$ and $\underline{a} = 1.85$ are practically equal to the values for coagulation and for the counter ion

adsorption (7), one can consider the present results as a support for the ideas on the interconnection between constituent I$^-$ ion adsorption, counter ion adsorption and coagulation.

It follows from fig. 3 that the coagulation values marked by arrows practically coincide with the intersections of the tangents through the steep parts of the plots and the horizontal tangents through the minimum values. The somewhat greater deviation of both points for Na$^+$ is probably caused by the high NaNO$_3$ concentration causing activity factor corrections of concentrations and also inducing partial recrystallizations accompanying the coagulation. The coagulation values marked with arrows were obtained from the "10 minutes intensity of scattered light against logarithm counter ion concentration" plots of freshly prepared sols of low concentration (8). The plots have a steep part and the intersection of the tangent on the steep part with the abscissa represents the coagulation value. This means that the steep parts of the light scattering plots and the minimum adsorption capacities belong to the same systems in which the counter ion concentrations are higher than the coagulation values. The low intensity of scattered light plots belongs to stable systems having high or decreasing adsorption capacities. The coagulation value determined in the described way is therefore a discrimination concentration between the systems which do coagulate and those in which aging processes take place and are considered colloidally stable. The aging processes are dissolutions of particles which were formed immediately after the mixing and formations of new particles which are finally greater by orders of magnitude from the first formed particles. The range of colloidal stability, limited by the coagulation value is characterized therefore by adsorption capacities which are higher than the adsorption capacities in the range of coagulation. The coagulation value obtained in the described way is a limiting concentration not only between stability-coagulation ranges but also between the ranges of counter ion or constituent ion concentrations of high and of minimum adsorption capacities.

Summary

The plots „amount I$^-$ adsorbed against pI" on freshly precipitated AgI, mol per mol, are linear. At pI $= 8$ and at pI$_{\text{limit}} = 9.4 - 9.1$ abrupt changes of the slopes can be observed. The slopes of the plots, i.e. the adsorption capacities are proportional to the logarithm of the starting I^- concentration and have common zero points of adsorption. The initial I^- concentration determines the number of primary particles formed and the number of I^- adsorbed per primary particle, i.e. at the starting I^- concentration the number of I^- adsorbed is maximal and at the Ag^+ concentration at which the isoelectric coagulation sets in the number of I^- adsorbed is minimal. The slope of the plots is explained as being proportional to the difference between the maximal and minimal number of I^- adsorbed and to the number of primary particles. The slopes of the plots are the higher the lower is the sol concentration and they are the lower the longer were the times elapsed between the precipitation and the measurement. The presence of cations of various valencies in a concentration higher than the critical one for coagulation cause the minimal adsorption capacity. The increase of the cation valency for 1 causes a shift to lower values of the critical concentration which shift is practically equal for the Na^+-Ba^{2+} and Ba^2-La^{3+} pairs.

Zusammenfassung

Die Funktionen „adsorbierte J^--Menge pJ" von frischem Silberjodid, Mol per Mol, sind linear. Bei $pJ = 8$ und $pJ_{Grenze} = 9.4 - 9.1$ werden plötzliche Änderungen der Neigungen beobachtet. Die Neigungen der Geraden, d.h. die Adsorptionskapazitäten sind dem Logarithmus der Anfangskonzentration der J-Ionen proportional und haben einen gemeinsamen Nullpunkt der Adsorption. Die Anfangskonzentration der J-Ionen bestimmt die Zahl der gebildeten Primärteilchen und die Zahl der adsorbierten J-Ionen per Primärteilchen. Das heißt bei der Anfangskonzentration der J-Ionen ist die Zahl der adsorbierten J-Ionen maximal, und bei der Ag-Ionenkonzentration, bei welcher die „isoelektrische" Koagulation eingeleitet (pJ_{Grenze}) wird, ist die Zahl der adsorbierten J-Ionen minimal. Die Neigung der Geraden wird durch die Proportionalität mit der Differenz zwischen der maximalen und minimalen Zahl der adsorbierten J-Ionen und mit der Zahl der gebildeten Primärteilchen gedeutet. Die Neigung der Geraden ist je höher, je niedriger die AgJ-Konzentration ist, und ist je niedriger, je länger die Zeit zwischen der Zubereitung des Sols und der Messung war. Die Anwesenheit von Kationen verschiedener Wertung in Konzentrationen, die höher sind als die Koagulationskonzentration, verursacht die niedrige Adsorptionskapazität. Eine Erhöhung der Wertigkeit des Kations um eins verursacht eine Verschiebung der Kationenkonzentration mit konstanter Kapazität zu niedrigeren Werten. Die Verschiebung im logarithmischen Maßstab ist praktisch gleich bei den Na^+/Ba^{2+} und Ba^{2+}/La^{3+} Kationenpaaren.

References

1) *Vincent, B., B.H. Bijsterbosch*, and *J. Lyklema*, J. Colloid Interface Sci. **37**, 171 (1971).
2) *Mirnik, M., B. Težak*, Trans. Faraday Soc. **50**, 65 (1954).
3) *Horne, O., E. Matijević, R. H. Ottewill, C. F. Weymouth*, and *M. C. Rastogi*, Trans. Faraday Soc. **161**, 50 (1958).
4) *Despotović, R., Z. Despotović, M. Mirnik*, and *B. Subotić*, Croat. Chem. Acta **42**, 557 (1970).
5) *Despotović, R.* and *B. Subotić*, Croat. Chem. Acta **43**, 153 (1971).
6) *Tesla-Tokmanovski, D., M. J. Herak, V. Pravdić*, and *M. Mirnik*, Croat. Chem. Acta **37**, 79 (1965).
7) *Herak, M. J.* and *M. Mirnik*, Kolloid-Z. **179**, 130 (1961).
8) *Težak, B., E. Matijević*, and *K. Schulz*, J. Phys. Colloid Chem. **55**, 1557, 1567 (1951).

Authors' address:

M. Mirnik
Laboratory of Physical Chemistry, Faculty of Sciences, University of Zagreb
41001 Zagreb (Yugoslavia)

S. Musić
"Ruder Boskovic" Institute, Zagreb
P.O.B. 1016
41001 Zagreb (Yugoslavia)

Progr. Colloid & Polymer Sci. **61**, 46–53 (1976)
ISSN 0340-255 X

Plenary lecture of the IUPAC-Conference on Colloid and Surface Science in
Budapest, September 15–20, 1975

National Academy of Sciences, Washington, D.C. (U.S.A.)

Clay-water relationship — theory and application. A review *)

H. van Olphen

With 4 figures

(Received December 9, 1975)

Introduction

A swelling clay can lift a building, a seemingly
solid soil may suddenly yield and cause a land
slide, clays form gels at relatively low concentrations, but gels and even concentrated pastes
can be changed to a free flowing mass by the
addition of very small amounts of chemicals.
Such pronounced physical effects are typical for
clay-water systems. Therefore, it has been
thought that the interaction energy between
clay and water is exceptionally large, and is also
of a long range character. However, on a unit
area basis, the adsorption energy of water on
clay surfaces is of the same order as that for
other inorganic surfaces. Also, adsorption isotherms show that the adsorption energy decays
rapidly with distance from the clay surface
within the first few layers of water adsorbed.
The special feature of clay particles is their
plate-like shape, and the large bulk effects can
be explained satisfactorily by considering particle to particle interactions, without having to
resort to assumption regarding long range effects
of the clay surfaces on the structure of the water
phase. The existence of such long range effects is
still a matter of controversy, but such effects,
(or even short range effects which certainly
exist) do not play any significant role in the
above phenomena. The short range effects (say
up to 1 nm from the surface) are, however,
important in phenomena involving the first few
layers of adsorbed water, for example when
dealing with the forces required to remove
adsorbed water from shales by overburden
pressure in geological formations.

---

*) National Academy of Sciences, Washington D.C.
(Author's address).

We shall discuss the theoretical basis of the
bulk effects of particle to particle interaction,
both qualitatively and quantitatively.

The electrical double layers on clay particle surfaces

Most clays occur as plate-like particles, and
these plates consist of stacked single "unit"
layers, each consisting of tetrahedral silica
sheets, and octahedral alumina sheets sharing
the apex oxygen atoms of the tetrahedrons. In
kaolinites, unit layers consist of one silica and
one alumina sheet, in montmorillonites and
illites they consist of one alumina sheet and
two silica sheets, one on each side of the alumina
sheet. In the packages of unit layers, forming a
particle, the unit layers are held together by
van der Waals forces, and by electrical forces,
and also, in the case of kaolinites, by hydrogen
bonding. In montmorillonites unit layers can
be separated by adsorption of water and other
polar molecules ("intracrystalline swelling"), as
shown by an increase of the 001 spacing.

Within the unit layer lattices isomorphous
substitutions occur, such as substitutions of
four-valent silicon by three-valent aluminum in
the silica sheet, and of three-valent aluminum
by two-valent magnesium in the alumina sheet.
Consequently, the lattice has an excess of negative charge. This excess charge is compensated
by the adsorption of cations on the exterior
surfaces of the unit layers, or of the stack of unit
layers (1). The ions which compensate the
lattice charge remain associated with the clay
during drying and wetting. When the particles
are immersed in water, the cations on the
exterior surface assume an atmospheric distribution and a diffuse electrical double layer is

created on these exterior layer surfaces. Since the charge is determined by the lattice substitutions, the double layer is a constant charge double layer, i.e. unlike the double layers on most inorganic particles, the charge is independent of the composition of the medium, and the potential varies with the addition of "indifferent" electrolytes.

The charge compensating cations in montmorillonites and illites are located both on the exterior surfaces of the packages and in between the unit layers in the stack. In kaolinites they are exclusively located on the exterior surfaces. All cations on exterior surfaces, the counter ions of the diffuse electrical double layers, are exchangeable. The cations located between the unit layers in montmorillonite are also exchangeable, but not those in the nonexpandable illites. For montmorillonites, therefore, the charge density of the unit layer surfaces can be calculated from the cation exchange capacity and the dimensions and weight of the unit cell. For illites and kaolinites the surface area of the particles must be determined in order to calculate the surface charge density from the cation exchange capacity. On a unit weight basis the cation exchange capacity of montmorillonite is rather high, of the order of 100 meq/100 g of clay, but rather low for kaolinites consisting of rather thick particles, e.g. 1—10 meq/100 g. Nevertheless, the density of surface charge of kaolinites is often higher than that of montmorillonites[1]).

The face surfaces of the unit layers or plates are not the only surfaces which are exposed. The edge surfaces, though relatively small in area, should also be considered. At these edge surfaces the crystal lattices of the particles are disrupted and bonds are broken in the silica and alumina sheets. At such broken bond surfaces an electrical double layer is created by the specific preferential adsorption of certain ions — the potential determining ions. The formation of an electrical double layer at the edge surfaces may be expected to correspond with that on the surfaces of alumina particles and of silica particles. For alumina particles a positive double layer is formed in an acid environment, and a negative double layer in alkaline suspensions.

At some intermediate pH of the solution, the surface is uncharged. The position of the point of zero charge depends on the crystal structure of the alumina (2). The surface of silica particles may be considered a model for the part of the edge surfaces at which the silica sheets are disrupted. Although silica particles are normally negatively charged, they do obtain a positive charge in the presence of very small amounts of aluminum ions in solution. Because of the solubility of the clay particles, aluminum ions are always present in the equilibrium solution, and therefore the broken silica sheets also may obtain a positive charge. Also, silica sheets may well be disrupted at sites where Si has been substituted by Al, and therefore, the edge surface may be essentially an alumina surface.

Therefore, the possibility should be considered that in a neutral or acid clay suspension the edge surfaces carry a positive double layer, although the electrophoretically determined charge is negative, possibly because of the dominating effect of the negative charge on the large expanse of face surface area (3), (4). Indeed, this concept is supported by several observations. For example, clay suspensions display a small anion exchange capacity which may be attributed to the exchange of the counter ions of a positive edge double layer (5). Another elegant support is the observed preferential adsorption of negative gold particles at the edges of kaolinite crystallites, as shown by electron micrographs of mixtures of suspensions of gold and kaolinite. This observation was first made by *Thiessen* (7). However, he interpreted this as a demonstration of the greater adsorption energy of edges and corners of crystals in general. The interpretation in terms of a positive edge charge is more acceptable since the preferential adsorption of gold particles is lost when the charge of the edge is reversed.

On the basis of the proposed dual character of the electrical double layer structure of clay particles, the flocculation and deflocculation behavior of clay suspensions can be readily understood, and the consequences of this behavior for the technologically important properties of clay-water systems can be analyzed. From such understanding the properties of clay-water systems can be tailored to suit any particular application through small adjustments in the composition of the water phase which affect the properties of the two double layers.

---

[1]) Isomorphous substitution as the origin of the surface charge of kaolinites has recently been questioned by *Ferris* and *Jepson* (6).

Colloidal stability and modes of particle association in clay suspensions

In suspensions of plate-like particles three different modes of particle association may occur: edge to edge, edge to face, and face to face. The consequences of these three types of association for the bulk properties of the suspensions may be expected to be quite different. Face to face association merely leads to the formation of thicker plates, and thus to a somewhat coarser suspension. Edge to edge and edge to face associations on the other hand, lead to the creation of a three dimensional card-house type matrix of interlinked particles, which will impart rigidity to the suspension if the matrix fills the entire available volume. In more dilute suspensions the linked particles will have the character of flocs. In the clay literature it has become customary to refer to flocculation and deflocculation only when edge to edge or edge to face association is involved, whereas face to face association is referred to as "aggregation" or "parallel aggregation", and the reverse process is called "dispersion".

The flocculation behavior of clay suspensions is rather complicated due to the variety of particle linking geometries. Moreover, the different modes of linking are governed by different sets of interparticle forces: The summation of van der Waals forces leads to different total forces of attraction, and since the double layers on edge and face surfaces are different, also the total electrical double interaction forces are different for the three modes of association. Under given conditions all three modes of association may occur simultaneously,

or one or another is dominant, and the relative importance of each mode of association can differ for different types of clays under the same set of conditions.

Since clay suspensions are flocculated more strongly with calcium ions than with sodium ions, the flocculation process must be dominated by the negative face double layer properties. The course of events in the flocculation process can best be studied from the changes in rheological behavior of the systems.

Colloidal stability and rheological behavior of clay-water systems

Dilute clay suspensions show Newtonian behavior, and concentrated systems display Bingham-type non-Newtonian behavior. The various modes of particle association will be reflected in the rheological behavior as follows: Edge to face and edge to edge association (flocculation) results in an increase of the viscosity because of the increased relative volume of the particles in the floc state in which they include water. In concentrated systems the linked matrix of particles results in the development of a yield stress. Disruption of the particle links (deflocculation) has the opposite effects. Face to face association results in a decrease of the viscosity of dilute suspensions which is proportional to the diameter to thickness ratio of the plates. In concentrated systems the yield stress is reduced because of the reduction of the number of particles, and hence the number of links in the matrix of edge to face and edge to edge linked particles.

As an example of the complex flocculation and rheological behavior of clay-water systems, we give an interpretation of data obtained for sodium-montmorillonite suspensions, based on the assumption that the edge double layers are indeed positive (8, p. 101 ff.). Figs. 1 and 2 show the relative viscosity of dilute and the Bingham yield stress of concentrated suspensions as a function of NaCl concentration. When dispersed in water, opposite charges on the clay surfaces result in edge to face association (9). In dilute systems such linking of the plates may be limited to T units or I units and since these would be very light units the suspension does not obtain a flocculated appearance. However, the formation of gels at higher concentrations indicates that the systems are indeed flocculated

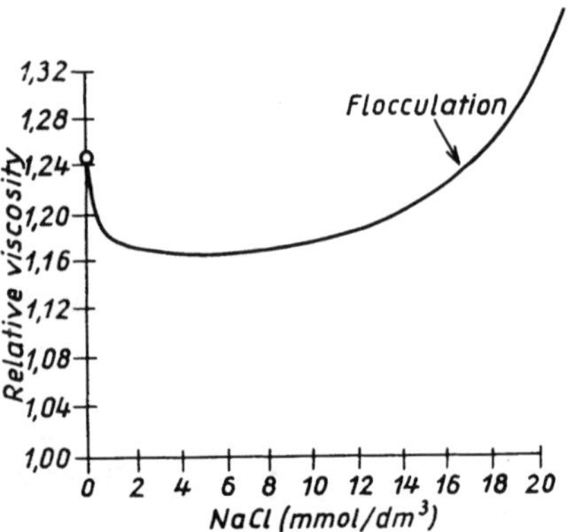

Fig. 1. Relative viscosity of a 0.23 percent sodium monmorillonite suspension as a function of the amount of NaCl added

in this stage, and colloidally unstable. Considering a single *I* unit, or such a unit as a section of the more or less cubic house of cards in the gel, there will be a net repulsive force between the two parallel plates (repulsion between the face double layers minus the van der Waals attraction between the plates), and a net attractive force between the connecting plate and the two parallel plates (electrostatic edge to face attraction plus van der Waals attraction). In the *I* unit, the connecting plate acts as a cross link (fig. 3). Apparently, in the electrolyte-free clay-water system, the attraction is larger than the repulsion. Upon the addition of a very small amount of NaCl, a considerable decrease of the yield stress is observed for the concentrated system, as well as some decrease of the viscosity in the dilute suspensions. This decrease may be explained as follows: The compression of both the face and the edge double layers in the presence of salt will cause a more effective screening of the surface charges, and hence the electrostatic edge to face attraction decreases. In the *I* units, the repulsion between the parallel plates — although somewhat reduced also — is now more powerful than the cross link and the structure breaks down resulting in a decrease of yield stress or viscosity.

With increasing amounts of NaCl, both yield stress and viscosity increase, and in the dilute suspensions flocculation is observed. With the further compression of the double layers, van der Waals attraction now dominates and both edge to face and edge to edge association is promoted. Also, face to face association becomes more important resulting in a gradual decrease of the yield stress of the concentrated system at very high electrolyte concentrations (not shown in fig. 2).

For different clays, the relative effects of various modes of association differ. For example, for kaolinite (fig. 2), after the initial decrease of the yield stress with small amounts of salt, further addition of salt does not cause significant changes of the yield stress. Apparently, face to face association is predominant in this system.

Deflocculation of clay-water systems

As long as opposite charges are present on the faces and edges of the clay particles, complete deflocculation of these systems cannot be achieved. In order to deflocculate the system

either the face charge or the edge charge must be reversed. Such reversal of charge can be achieved by the adsorption of certain ions. The reversal of edge charge by anions is practised most frequently since only small amounts of additives

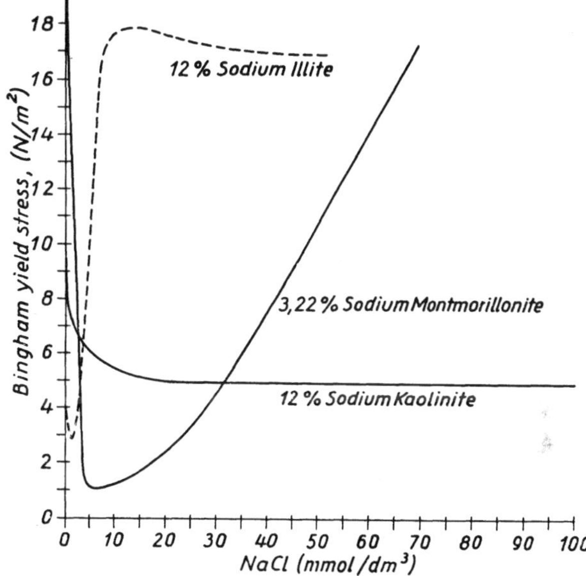

Fig. 2. Bingham yield stress of clay suspensions as a function of the amount of NaCl added

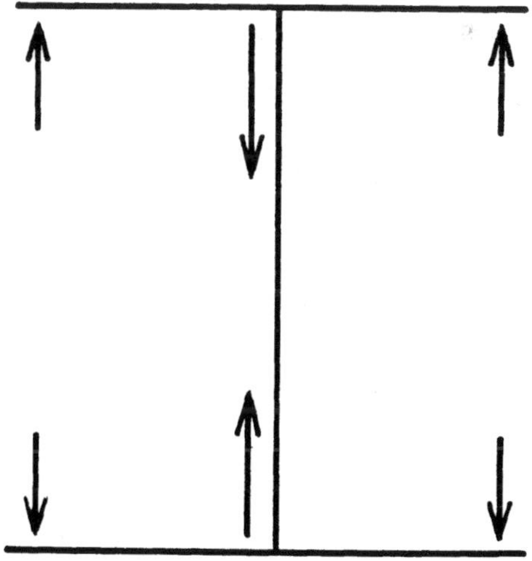

Fig. 3. *I*-unit of clay plates

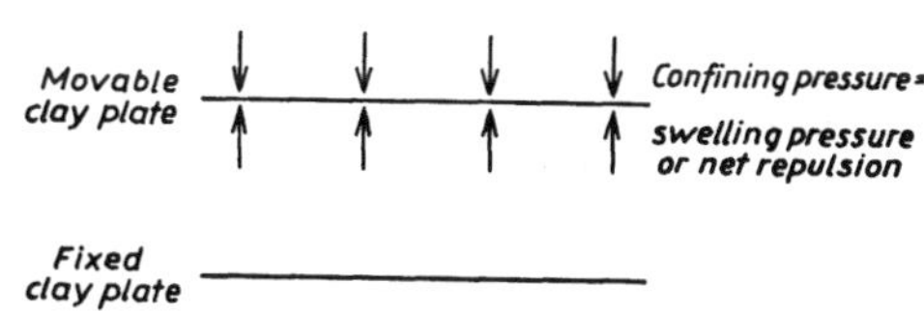

Fig. 4. Swelling forces in clay-water systems between parallel plates

are required to cover the relatively small edge surface areas. Anions which are able to reverse the positive edge charge are those which form complex anions with Al. They are chemisorbed in excess at the edges and the complex anion becomes the ion which constitutes the surface charge, and the cations of the added salt — usually Na ions — become the counter ions of the new negative double layer. Examples of suitable anions are pyrophosphate or meta-phosphate ions, organic anions of polyphenolic character (e.g. tannates), and fluorides. The charge can also be reversed by raising the pH of the system. The breakdown of the matrix upon the addition of very small amounts of these anions is evident from a dramatic thinning effect. Rheological measurements show that the thinning action is primarily the result of a decrease of the Bingham yield stress whereas the differential viscosity (the slope of the curve in the rate of shear-shearing stress diagram) changes little. An even greater reduction of the yield stress can be achieved by simultaneous addition of cations of higher valency which cause face to face association (or "aggregation"). This technique is often used in conjunction with treatment by tannates, by addition of lime to the system.

The sensitivity of the clay-water systems to additions of electrolytes which either flocculate or deflocculate the systems allows the tailoring of the properties of the systems to suit a variety of applications on an economical basis, such as the treatment of concentrated clay base drilling fluids or of ceramic slips, in order to give them the desired flow properties. On the other hand, this sensitivity sometimes has very undesirable consequences, for example on the mechanical stability of soils. Landslides may be caused by a breakdown of the matrix in an originally salty flocculated coherent soil when the salt is removed gradually by leaching, or by the adsorption of organic anions leached in the soil from surface vegetation by rain water. When attractive forces between particles in the matrix are gradually reduced and become repulsive, the soil may remain solid for a while since the clay plates support each other mechanically as in a house of cards. However, any slight disturbance will cause the matrix to collapse, and a landslide occurs suddenly. In such areas, the disturbed soil remains fluid, indicating that it is indeed in the colloidally stable condition with repulsive forces prevailing.

Quantitative evaluation of particle interaction from rheological data

As shown above, rheological observations are very helpful in interpreting changes in colloidal stability in these systems of clay, water, and electrolyte. In addition, rheological data can also be used to obtain quantitative information about particle interaction forces in these systems on the basis of reasonable models for the particle matrix, and information on the size and shape of individual particles. Such quantitative studies were made on montmorillonite gels, using well fractionated and homoionic clay for which average particle sizes were determined (10, 11). Referring to fig. 2, two regions should be distinguished in which gels are obtained, one in pure water, the other at high electrolyte concentrations. The gels in pure water are non-thixotropic, those at high electrolyte concentrations are strongly thixotropic. In a thixotropic system, stirring causes a thinning of the system, and the original yield stress is only slowly restored after stirring is discontinued. After breaking the links in the matrix by mechanical forces, restoration of the links is a slow process, analogous to slow coagulation in dilute systems. Apparently, there is an energy barrier for particle association in the thixotropic systems. In the pure systems which are nonthixotropic such a barrier is absent, as one would indeed expect from the concept that the gel is formed by positive edge to negative face interaction in this condition.

For the nonthixotropic gels in pure water, the matrix may be visualized as a more or less cubic card house of individual plates. The minimum clay concentration at which a continuous card house is created throughout the available volume can be computed from the size and shape of the particles. This computed minimum concentration appears to be equal to that at which a yield stress is observed, i.e. at only a few percent of clay in the case of the montmorillonite studied when in the sodium form. The particles of the same montmorillonite in the calcium form are about three to four times thicker than those in the sodium form. Accordingly, the minimum clay concentration for the development of a yield stress was observed to be three to four times larger for the calcium clay. These observations support the model of the cubic card house. On the basis of this model, the average linking force between the particles in the matrix can be computed, applying Goodeve's analysis (12) to the rheological data. *Goodeve* analyzed the Bingham flow behavior as follows: In the linear region there are two contributions to the shearing stress, one is a viscosity contribution which is proportional to the rate of shear, the other is a shear-independent contribution which is equal to the Bingham yield stress. The latter contribution can be interpreted as follows: Consider two linked particles situated in adjacent shear planes. Initially, during shear, the linked particles stretch and an elastic force is generated within the particles. As soon as this force becomes larger than the strength of the link, the link breaks. A moment later either particle engages with another one and a link is reestablished. During the process of stretching of the particles and breaking of the link, an impulse is transmitted from one shearing layer to the adjacent one. This impulse is inversily proportional to the rate of shear. The number of impulses transmitted per second per unit area is proportional to the rate of shear. Hence, the shearing force, which is the product of the magnitude of the impulse and the number of impulses, is independent of the rate of shear. The mechanism of shear making and breaking of links therefore explains the shear independent contribution, the Bingham yield stress. In the region of non-linearity of the Bingham flow curve, at low shear rates, thermal making and breaking of links cannot be neglected with respect to shear making and breaking of links. Applying this analysis to the actual rheological data for the sodium montmorillonite gel in pure water, an average particle linking force of the order of 10^{-9} N (10^{-4} dyne) was computed.

For the thixotropic systems at high salt concentration, the size of the energy barrier can be evaluated from measurements of the rate of recovery of the links, by applying Fuchs' theory for slow coagulation of colloidal systems (13). The problem of making such measurements in thixotropic systems is that the system is disturbed by the measurement. The following method allows measurement of thixotropic recovery rates without disturbing the system: From the rate of propagation of a very small shear wave which only causes elastic deformations without breaking links, the shear modulus of elasticity of the system can be determined as a function of time of recovery. The modulus of elasticity is a measure of the number of links which have been reestablished at the moment the measurement is made. The energy barriers in the systems studied, expressed in units of kT, were of the order of what is expected for slowly coagulating systems.

Swelling and compaction of clay-water systems

The swelling of clays in contact with water is often rather spectacular. Large increases in volume are observed and high pressures are developed when the system is confined. In a compacted clay there will be a tendency for the plates to become more or less aligned in a parallel fashion. We shall consider this arrangement first, and later discuss the effect of the presence of non parallel particles. Fig. 3 schematically shows a reference clay plate and a parallel movable plate above, in an environment of water. If under a given condition of particle distance (or water content) the clay has a tendency to swell, a net repulsive force must exist between the plates. This net repulsive force can be measured by applying a confining pressure to the upper plate which will keep the plates at the given distance (or which keeps the water content constant).

In an analysis of the nature and magnitude of the forces between the plates two different situations must be distinguished: one at which the plates are very close together, i.e. not more than about 1 nm apart, as in deeply buried shales, and another at which the plates are further apart as in surface clays. The long range forces which operate between the plates in the latter situation are the electrical double layer repulsion and the van der Waals attraction. In most situations in practice, the van der Waals attraction may be neglected with respect to the double layer repulsion, hence, the confining pressure will be approximately equal to the electrical double layer repulsion. In many cases, the double layer repulsion as measured from the confining pressure agrees well with that computed from electrical double layer theory, knowing the charge density of the particles. Experimental techniques and computations have been published by *Bolt* and *Miller* (14), *Warkentin, Bolt, and Miller* (15), *Warkentin* and *Schofield* (16). The swelling pressure due to the double layer repulsion can be appreciable, and often amounts to several atmospheres, enough to lift buildings.

Since the double layer repulsion between the plates may be considered an osmotic effect resulting from the difference in ion concentration between the plates and that outside, this type of swelling is often referred to as osmotic swelling. In some cases the double layer repulsion as measured by the confining pressure, does not agree with the calculated double layer repulsion. One reason for this discrepancy may be the presence of nonparallel particles which form cross links between the parallel particles and thus limit the swelling. From the evaluation of the strength of such links (see above) it can be shown that only very few such nonparallel particles could account for the lower swelling pressures (17, 18).

The volume changes accompanying osmotic swelling may amount to as much as a factor 10 to 100 with reference to the dry clay. The volume change is greater when the individual particles are thinner, hence the increase in volume (and thus the movement of a building on a swelling clay) is more pronounced for montmorillonites and some fine grained illites than for coarser clays.

Although all clays show osmotic swelling, the clays of the montmorillonite group are often specifically referred to as the swelling clays, since they show interlayer swelling. The intercalation of water is usually limited to at most four monomolecular layers of water, amounting to a doubling of the volume. In the discussion of this short range swelling, it is convenient to consider the parting of the unit layers in montmorillonite crystallites, since the effect can be measured by X-ray techniques. However, the analysis of the intracrystalline swelling of montmorillonites applies equally to the short range interaction of individual particles of any type of clay during compaction and swelling processes.

The forces operating in the range of very small particle separations are threefold: the van der Waals attraction, the electrical interaction, and the adsorption forces between clay and water molecules. In this situation, the electrical interaction is probably attraction rather than repulsion, since at small plate separations, the counter ions are usually (but not always) located midway between the negatively charged plates. The adsorption forces between clay and water manifest themselves as a repulsion since during approach of the plates the removal of

adsorbed water requires work. In general, the net work which must be done by compaction forces equals the work of desorption of water minus the work due to van der Waals and electrical attraction forces. When contacting the dry clay with water, the reverse processes take place and the adsorption of water is the main cause of the short range swelling. Using the intracrystalline swelling of montmorillonite clays as a model for the short range swelling of clays in general, the magnitude of the net swelling pressure can be computed from water vapor adsorption isotherms and X-ray data for the unit layer separation as a function of vapor pressure. For example, from adsorption isotherms for a vermiculite clay, the work required to remove the last monomolecular layer of water was found to be about 10 J/m^2 (100 erg/cm^2). X-ray data show that the plates move a distance of 0.25 nm during desorption with respect to each other. Hence, the compaction pressure (or swelling pressure) in this region amounts to $10/0.25 \times 10^{-9} = 4 \times 10^8 \text{ N/m}^2$ or 4000 bar (19), (20). These computations (which as to order of magnitude could be confirmed by direct measurement) indicate that commonly overburden pressure will be insufficient to remove the last layers of water in deeply buried shales.

Several studies have been made of the properties of the interlayer water in montmorillonite type clays. For example, for a highly charged vermiculite, the entropy of the adsorbed water as calculated from adsorption isotherms and calorimetric data was found to be lower than that of bulk water, indicating a decreased mobility of the water molecules in interlayer position (20). The restricted mobility of the water in this system was confirmed by infrared and NMR investigation of the interlayer hydrates (21). X-ray data on montmorillonite hydrates also indicate that the adsorption of water affects the solid substrate, as shown by small changes in the b-dimensions of the lattice (22).

Summary

The magnitude of colloidal and mechanical effects of the interaction of clays and water is unusually large, therefore, long range effects of surface forces on the structure of water are often assumed to exist in these systems. However, such effects can be readily explained on the basis of electrical and dispersion forces between the plate-like clay particles, according to a qualitative and quantitative analysis of such forces, and in particular the forces due to the interaction of electrical double layers. Applications in such diverse areas as foundation engineering, land slide analysis, oil drilling fluid management, and ceramics production are briefly discussed as illustrations.

Short range interaction — up to about 1 nm — between clay and water does affect the structure of water which is adsorbed on the clay particle surfaces, as is shown by the evaluation of the entropy of adsorption and from IR and NMR studies of water adsorbed by expanding clays. An illustrative application of the quantitative analysis of adsorption forces is an evaluation of the retention of water in shales under overburden pressure.

References

1) *Marshall, C. E.*, Z. Krist. **91 A**, 433 (1935).
2) *van Schuylenborgh, J.*, Rec. Trav. Chim. Pays-Bas **70**, 985 (1951).
3) *van Olphen, H.*, Rec. Trav. Chim. Pays-Bas **69**, 1308, 1313 (1950).
4) *van Olphen, H.*, Disc. Far. Soc. **11**, 82 (1951).
5) *Schofield, R. K.* and *H. R. Samson*, Disc. Far. Soc. **18**, 135, 220 (1954).
6) *Ferris, A. P.* and *W. B. Jepson*, J. Colloid Interface Sci. **51**, 245 (1975).
7) *Thiessen, P. A.*, Z. Elektrochem. **48**, 675 (1942).
8) *van Olphen, H.*, An Introduction to Clay Colloid Chemistry (New York 1963).
9) *van Olphen, H.*, J. Colloid Sci. **19**, 313 (1964).
10) *van Olphen, H.*, Clays, Clay Min. **4**, 204 (1956).
11) *van Olphen, H.*, Clays, Clay Min. **6**, 196 (1959).
12) *Goodeve, C. F.*, Trans. Faraday Soc. **35**, 342 (1939).
13) *Fuchs, N.*, Z. Physik **89**, 736 (1934).
14) *Bolt, G. H.* and *R. D. Miller*, Soil Sci. Soc. Amer. Proc. **19**, 285 (1955).
15) *Warkentin, B. P., G. H. Bolt*, and *R. D. Miller*, Soil Sci. Soc. Amer. Proc. **21**, 495 (1957).
16) *Warkentin, B. P.* and *R. K. Schofield*, Clays, Clay Min. **7**, 343 (1960).
17) *Norrish, K.* and *J. A. Rausell Colom*, Clays, Clay Min. **10**, 123 (1963).
18) *van Olphen, H.*, J. Colloid Sci. **17**, 660 (1962).
19) *van Olphen, H.*, Clays, Clay Min. **11**, 178 (1963).
20) *van Olphen, H.*, J. Colloid Sci. **20**, 822 (1965).
21) *Hougardy, J., J. M. Serratosa, W. Stone*, and *H. van Olphen*, Special Disc. Far. Soc. **1**, 187 (1970).
22) *Ravina, I.* and *P. F. Low*, Clays, Clay Min. **20**, 109 (1975).

Author's address

H. van Olphen
National Academy of Sciences
2101 Constitution Avenue, N.W.
Washington, D.C. (USA)

Progr. Colloid & Polymer Sci. **61**, 54–63 (1976)
© 1976 by Dr. Dietrich Steinkopff Verlag GmbH & Co. KG, Darmstadt
ISSN 0340-255 X

Plenary lecture of the IUPAC-Conference on Colloid and Surface Science in
Budapest, September 15—20, 1975

Center for Surface and Coatings Research, Lehigh University, Bethlehem, Penna. (USA)

Water on silica and silicate surfaces
IV. Silane treated silicas

H. H. Hsing) and A. C. Zettlemoyer*

With 18 figures and 5 tables

(Received December 9, 1975)

A number of studies of silica surfaces have
emanated from this Laboratory. Assessment of
surface areas by both argon and nitrogen
adsorption has been examined (1); ice nucleation
by partially hydrophobed silicas has been studied
(2); water adsorption (3) and near infrared
reflectance (NIR) spectroscopy have also been
employed to study these silicas (4) which have
some of the surface hydroxyls removed by
thermal treatment (2). In these investigations
the parent material was HiSil 233, a wet-
precipitated silica, described further in table 1.
Partially hydrophobed products were compared
to a Cab-o-Sil, a flame hydrolyzed aerosil type,
possessing a lower surface concentration of
hydroxyls. The apparent too high density of
OH's on the HiSil (table 1) strongly suggests
the presence of micropores, a contention already
put forward by *Sing* (5). The ice nucleation
work had already suggested that the residual
OH's on Cab-o-Sil are widely separated. This
conclusion is supported by the work reported
here.

Table 1. Silica samples

HiSil 233	$123 \ m^2/g$
Wet precipitated	Ar: $16.0 \ A^2$
	1 OH/7 A^2
Cab-o-Sil M-5	$178 \ m^2/g$
Aerosil; $SiCl_4$	1 OH/40 A^2

We have pioneered the use of the NIR technique (4) to study water adsorption on silicas.
Fig. 1 displays a sketch of the apparatus. Since
a loose bed of the powder is used, the problem of
adherent particles and distorted surfaces as in
pressed discs is minimized. Further, the temperature control is far better (within 0.2 °C)
than in transmission infrared where the temperature rise can be over 50 °C. In addition, we
scan with monochromatic light to reduce the
heating effect. The addition of water vapor is
made through the obvious port provided for the
purpose. Fig. 2 shows the NIR spectra obtained
for one of the thermally treated HiSils. This
overtone region is rich in interesting bands while
lacking in background from the silica. At about
$7300 \ cm^{-1}$, the SiOH stretch overtone band ($2\,\nu$)
occurs. The addition of water diminishes this
band while increasing the neighboring H-bonded
OH band at $7150 \ cm^{-1}$. The band at $5300 \ cm^{-1}$
is a combination band ($\nu + \delta$) which develops
as the water molecules are added to the surface.

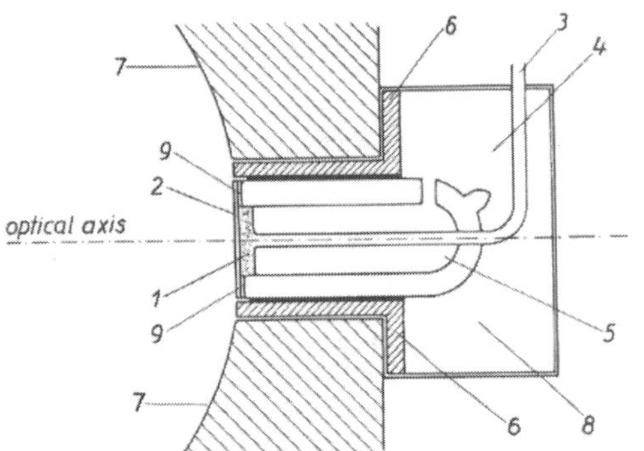

Fig. 1. Arrangement used to measure NIR absorption.
The bed of silica powder can be exposed to water or
other vapors so that the spectra can be followed as a
function of coverage

*) Ashland Chemical Company, Columbus, Ohio
43216.

Fig. 2. The spectra obtained as water is adsorbed on partially dehydrated HiSil. The No. 1 spectrum shows the absence of absorption at 5300 cm^{-1} because this case represents the absence of physically adsorbed water. This band develops as water is adsorbed, while the broad band at about 7150 cm^{-1} grows at the expense of the sharp peak at 7300 cm^{-1}

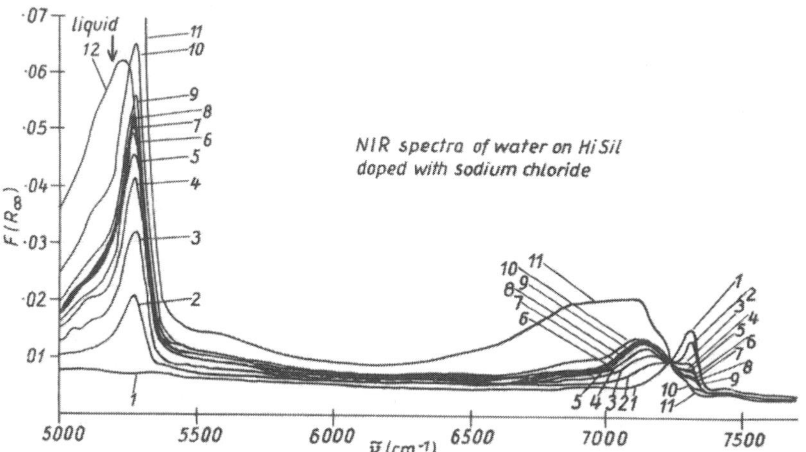

It was a natural consequence of our earlier work to attach organic ligands to the silicas and to study, by isotherms and by the NIR technique, the interaction of water with these treated silicas. The organosilanes employed in this study are depicted in table 2, along with the shorthand notation used hereafter. It turns out to be

Table 2. Organo silicon ligands

$(CH_3)_3SiNSi(CH_3)_3$	HMDS
$\quad$ H	
$CH_2=CH-SiCl_3$	$C=C-$
$HS(CH_2)_3Si(OCH_3)_3$	$HS-$
$CH_2-CHCH_2O(CH_2)_3Si(OCH_3)_3$	CH_2-CH-
$\quad\diagdown\ \diagup$	$\quad\diagdown\ \diagup$
$\quad\quad O$	$\quad\quad O$
$NH_2(CH_2)_3Si(OC_2H_5)_3$	NH_2-

fortunate that both the HiSil and the Cab-o-Sil were employed; the two treated surfaces in each case behave quite differently toward water and the alkoxylated surfaces are quite different. The HMDS breaks up upon reaction with the silica surface and the question has never been addressed in the past as to whether the resulting NH_3 interacts with the surface. The $Si(CH_3)_3$ groups occupy about 40 A^2, so there is just about one ligand for each OH group on Cab-o-Sil if the OH's are equispaced. Besides HMDS, only the unsaturated C=C compound is reacted from the vapor phase albeit at elevated temperatures. The other three compounds are reacted from solution as we shall see below. The vinyl compound and the three alkoxysilane compounds are used in the surface treatment of glass fibers used in polymer composites; therefore, there is great interest in water interaction with these treated surfaces. Furthermore, it

was thought that organic ligands fixed to a tractable surface such as silica would provide a more viable path to the study of their interaction with water than would cell membranes, water dispersions of the organics and the like.

No previous workers have examined water interaction with silane-treated silica. The interaction of HMDS was reported in 1968 by *Stark* et al. (5) and the kinetics of the reaction with an undefined silica by *Hertl* and *Hair* (6) in 1971. *Kiselev* and his co-workers (7) in 1961 and 1964 examined the reaction of trimethylchlorosilane with silica, as did *Synder* and *Ward* (8) in 1966, and *Hair* and *Hertl* (9) for all the methyl chlorosilanes. *Stark* et al. (5) in 1968 also examined the reaction of trimethylhydroxysilane, trimethylchlorosilane and HMDS with silica. *Lee* (10) in 1968 examined the product of hydroxysilane treatment of glass slides by contact angle measurements in an attempt to learn the configurations of the ligands on the surface. In 1970, *Kharilonov* et al. (11) also reported on their studies of hydroxysilane reactions with silicas.

HMDS-treated silicas. An advantage of HMDS is that it reacts with the S_s-OH's (S_s is a surface silicon atom) from the vapor state even at room temperature. The surface OH's are titrated by the HMDS according to:

$$HMDS + 2 \to S_s-OH \to$$
$$\to 2 \to Si_s-O-Si(CH_3)_3 + NH_3 . \quad [1]$$

The question as to whether the NH_3 also reacts with the surface hydroxyls had not been discussed heretofore. The cross-sectional area of the trimethyl ligand is about 40 A^2; therefore on Cab-o-Sil, if the sparse OH's are equispaced, each will react with HMDS (if little NH_3 is taken up). On the other hand, the surface density of OH's on HiSil is so high that many OH's remain after total ligand coverage; the question of

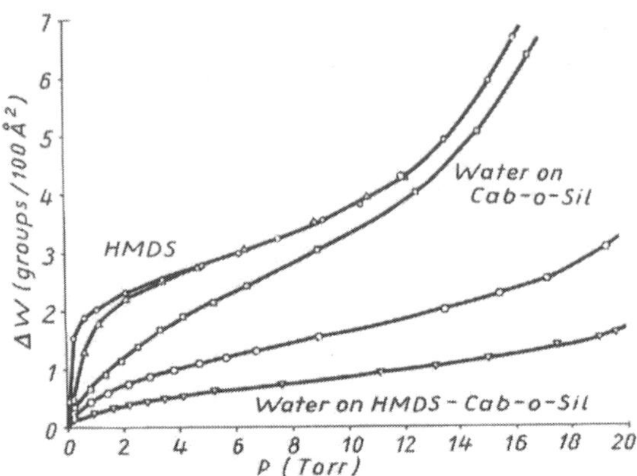

Fig. 3. The upper two curves are the HMDS isotherms for HiSil 233-o and for Cab-o-Sil M-5 △ at 25 °C. The lower three curves (discussed later) are water vapor adsorption isotherms for bare Cab-o-Sil □, and HMDS-treated Cab-o-Sil: 0.92 ligands/100 A² ◇ and 1.75 ligands/100 A² ▽, all at 25 °C. The latter coverage produces considerable hydrophobicity

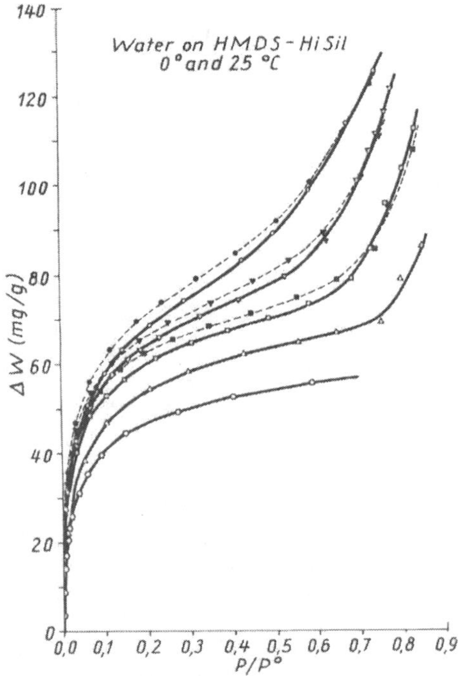

Fig. 4. Water vapor isotherms on HMDS-treated HiS at four coverages: 0% HMDS ○, 29.6% ▽, 52.2% ⊔ and 100% ◇ at 25 °C; solid points are for 0 °C. The lowest curves ◇ for 100% HMDS was determined earlier by a volumetric technique; the apparent 25% lower adsorption is believed to be due to experimental problems such as adsorption on the sensor

assessibility of the remaining OH's to water arises. These residual OH's might be regarded as being covered with an umbrella formed by the $-Si(CH_3)_3$ ligands.

Adsorption isotherms for HMDS on both the HiSil and the Cab-o-Sil surfaces at 25 °C in units of $-Si(CH_3)_3$ ligand groups/100 A² are plotted in Fig. 3. From both B points, the $Si_8-O-Si(CH_3)_3$ cross-sectional area is 40 A², in accord with the estimate from the atomic model. It is interesting that the B point for the initially fully hydroxylated HiSil is reached at a lower HMDS pressure than it is for the Cab-o-Sil, suggesting that the HiSil possesses the more reactive OH's. The fact that the B point for the initially 1 OH/40 A² populated Cab-o-Sil is the same as that for the HiSil indicates that the surface OH's on Cab-o-Sil must be almost equidistantly spaced.

Water vapor isotherms at 0 and 25 °C for HiSil are plotted in fig. 4 for different amounts of chemisorbed HMDS from 0 to 100%. Controlled amounts of HMDS were introduced and then the amounts retained were ascertained gravimetrically after exhaustive pumping at room temperature. The HMDS percentages are based on the coverage of one ligand per 40 A² as being 100%. The water isotherms are strongly Type II and retain a sharp knee even after high HMDS coverage. The organic ligand does not inhibit the first layer adsorption of water.

It is interesting to merge the HMDS and water vapor isotherms for HiSil as presented in fig. 5. They coincide in the monolayer region but

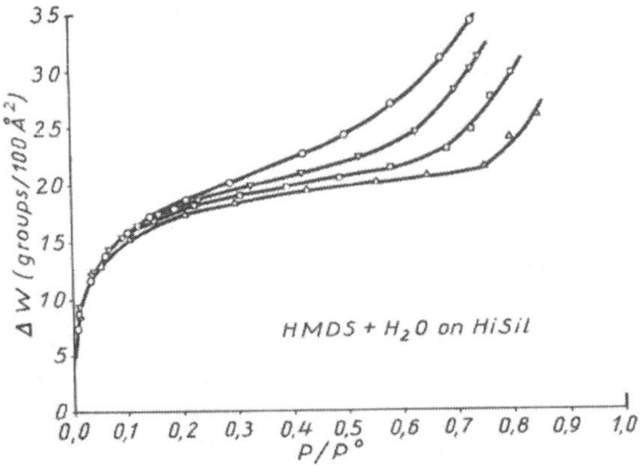

Fig. 5. Composite isotherms for water plus HMDS ligands on HiSil. The similar B points and repressed multilayer adsorption, especially at the higher ligand coverage, is expected. Composite monolayer values are also shown to agree in table 3

diverge at higher relative pressures as might be expected. The chemisorbed ligands restrict the growth of the second and subsequent layers, the more so at higher HMDS coverages. Table 3 gives these combined monolayer values based on the BET model. (We shall learn below that the NIR results make this model doubtful, but comparison between isotherms in this manner are still useful.) The resultant totals of H_2O + ligands are quite close. However, the molecular area for the water molecules is only about 7 A^2, considerably smaller than that reported from this laboratory heretofore (2). (The earlier result, it should be noted, was based on the more doubtful volumetric method.) This apparently small value for the co-area of water supports the contention of *Sing* (12) that HiSil contains a minor amount of micropores. While water would enter the micropores, of course the HMDS ligands would not.

Table 3. HMDS + H_2O on HMDS-HiSil

HMDS Groups/ 100 A^2	HMDS %	H_2O's/ 100 A^2	HMDS + H_2O/100 A^2
0	0	15.8	15.8
0.73	29.6	14.8	15.6
1.28	52.2	14.1	15.4
2.45	100	13.0	15.5

Isosteric heats for water adsorption at three different HMDS coverages on HiSil are presented in fig. 6. These curves overlap and only approach the heat of liquefaction at $\Theta = 2$. The interaction of water with the underlying OH's is independent of the HMDS coverages. These values are higher than have been reported at 0.5 to 1.0 Θ for partially dehydrated HiSil (1, 2). Micropores may explain the high heats and these may be eliminated by the heat treatments used to produce the partially dehydrated samples.

Water adsorption isotherms for HMDS-treated Cab-o-Sil are compared with that for the bare Cab-o-Sil in fig. 3. At the highest HMDS coverage after pumping to 10^6 torr at 25 °C, there are 1.75 ligands/100 A^2 and the remaining OH's yield a water monolayer value of 0.58 by the BET model (probably some of which is adsorbed on top of adsorbed NH_3; see below). The latter value has been corrected to about 0.7 from the NIR studies reported below. The total of 2.45/100 A^2 gives about 40 A^2 per

adsorbate unit as expected. This total is somewhat smaller, however, than the NIR monolayer value of 2.9 for water for the bare Cab-o-Sil reported below.

Examples of the NIR spectra obtained for HiSil at several increasing HMDS pressures are presented in fig. 7. The peak intensities of the several bands are plotted against HMDS pressure in fig. 8. The C—H stretch frequencies at 5730 and 5880 cm^{-1} produce curves similar to the adsorption isotherms as might be expected. The Si_sOH band (2ν) is shifted slightly from 7289 to 7241 cm^{-1} apparently due to perturbation by the adsorbed ligands; this band diminishes with pressure and then levels off above 2 torr in accord with the HMDS isotherm.

The band at 6530 cm^{-1} is assigned to the N—H stretch frequency. The development of this band parallels the diminution in the Si_sOH band. This evidence for the ammonia adsorption is, we believe, the first that has been

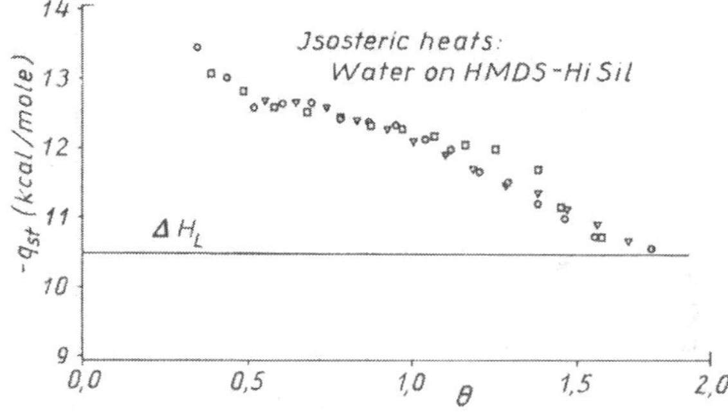

Fig. 6. Isosteric heats of adsorption on HMDS-treated HiSil: 0% ○, 29.6% ▽, 52.2% □, showing almost no effect of the presence of ligands on the energetics of water adsorption. Band assignments are indicated here and in table 2

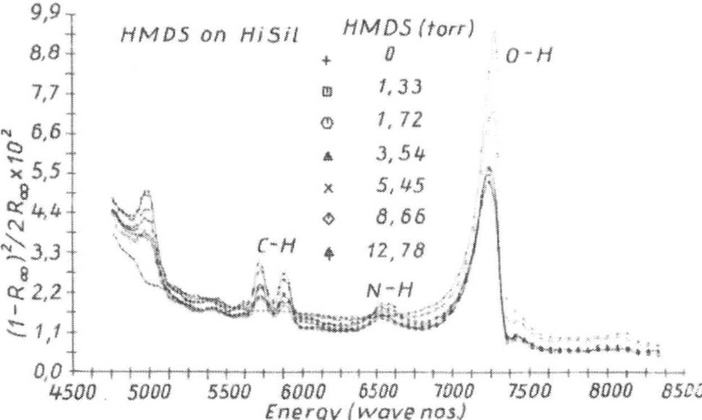

Fig. 7. Typical NIR reflectance spectra at a variety of pressures of HMDS adsorbed on HiSil at 25 °C

reported. The shoulder which develops at 6610 cm^{-1} is ascribed to N—H stretch in the build up of the second layer of physically adsorbed HMDS. Thus, the complete picture of chemisorbed and physisorbed HMDS on HiSil is established by this spectral analysis.

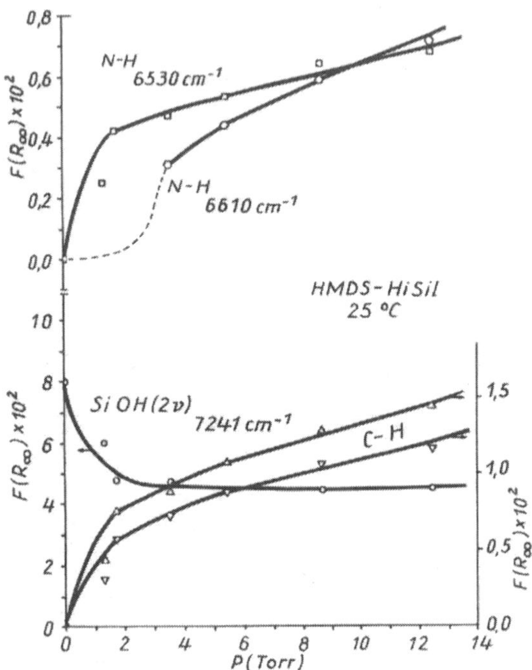

Fig. 8. Intensity of bands (Schuster-Kubelka-Monk function) versus HMDS pressure at 25 °C from fig. 5: ○ Si$_s$OH (2ν) band at 7289 cm^{-1}; △ and ▽ C—H bands at 57300 and 5880 cm^{-1}; □ N—H(NH$_3$) at 6530 cm^{-1} and ◇ N—H (HMDS) at 6610 cm^{-1}

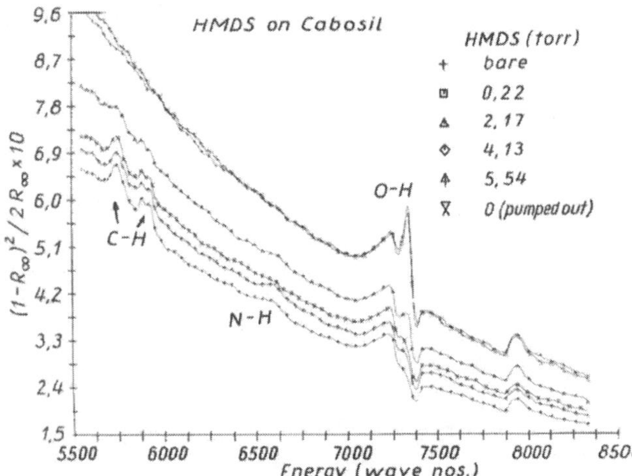

Fig. 9. Typical NIR reflectance spectra at a variety of HMDS pressures on Cab-o-Sil at 25 °C

The HMDS-treated HiSil was further studied by obtaining spectra after evacuating to 10^{-6} torr at room temperature and then at a sequence of elevated temperatures. Evacuation at room temperatures reduces all the bands except for a slight increase in the Si$_s$OH (2ν) band. The N—H(NH$_3$) band at 6530 cm^{-1} decreases only slightly indicating the strong adsorption of NH$_3$ to the surface OH's. Apparently, the main event upon evacuation at room temperature is the loss of physically adsorbed HMDS. Furthermore, little change in band intensities occurs up to 50 °C. Large changes then occur up to 140 °C which plateau at 170 °C. The complete elimination of the N—H(NH$_3$) band at 170 °C is of interest because the same temperature is required at 10^{-6} torr to remove physically adsorbed water from HiSil. The shift of some 80 cm^{-1} (6610 to 6530 cm^{-1}) suggests a strong perturbation in the decrease of the N—H vibration frequency from HMDS to NH$_3$ adsorption.

The NIR reflectance technique was also applied to the bare and HMDS-treated Cab-o-Sil surfaces. The spectra are displayed in fig. 9. The assignments made for the bands are tabulated in table 4 along with those for the HiSil. The most striking discovery is that two

Table 4. NIR bands: HMDS — Silicas

Freq. cm^{-1}	Surface	Group Assigments — Overtones
4975	Hisil & Cabosil	Adsorbed NH$_3$-bending + assymetric stretch
5730 & 5880	Hisil & Cabosil	C—H stretch
6530	Hisil	N—H stretch — chem. NH$_3$
6580	Cabosil	N—H stretch — chem. NH$_3$
6610	Hisil	N—H stretch — phys. NH$_3$
7246	Cabosil	O—H stretch — hydrogen bonded
7289	Hisil	O—H stretch (2ν) — free OH's
7331	Cabosil	O—H stretch — free OH's

distinct bands due to Si—OH are found at 7246 and 7331 cm^{-1}. In the transmission mode, only a mixed (3665 and 3749 cm^{-1}) band has been reported (13) with a wider shoulder on the low frequency side attributed to hydrogen bonded OH's or bulk OH groups. The results presented here indicate that hydrogen bonded OH's either do not react with HMDS as suggested previously (6), or, as seems more likely to us, that bulk OH's are responsible for the low frequency peak.

Additionally, the N—H(NH₃) is both weaker on the Cab-o-Sil surface and is totally removed by evacuation at room temperature; its appearance at a higher frequency (6580 cm⁻¹) also suggests weaker interaction. Furthermore, the Si_sOH band (7331 cm⁻¹) is almost entirely quenched again indicating the rather equal spacing of the OH groups on the Cab-o-Sil. The band intensities versus the pressure are plotted in fig. 10. In accord with the isotherms, these curves are not as sharp as for HiSil. The dashed lines refer to the changes on evacuation at room temperature. Both the C—H bands and the N—H band diminish on evacuation indicating weaker interaction with the Cab-o-Sil surface.

NIR reflectance spectra were also measured to follow water adsorption on the bare and HMDS-treated silica surfaces. Since similar spectra were reported on the bare surfaces heretofore (4), these spectra are not displayed here. As noted before, water adsorption on the HMDS-treated HiSil is little affected by the HMDS treatment except for a reduction in the multilayer region.

The intensity curve (SKM function) for the 100% HMDS-treated HiSil is given in fig. 11. The tailing off suggests clustering of the water molecules around molecules adsorbed at lower pressures (4). And the extrapolated value is believed to be a more appropriate measure of the residual surface OH concentration than is given by the BET model or B point value from the isotherm. Of course, the diminution of the Si_sOH (2ν) peak indicates that adsorption occurs on the OH's below the silazane ligands; apparently also about 50% of the OH's are titrated before clustering becomes important as indicated by deviation from the linear plot. Even at high water vapor relative pressures, the spectra indicate considerable difference of the adsorbed water from bulk water.

Water adsorption is too low on 100% HMDS-treated Cab-o-Sil to produce very useful results from the spectra. The intensity results on bare Cab-o-Sil are also plotted in fig. 11 for comparison. The extrapolated monolayer value is 15.9 mg/g (2.9 molecules/100 A²). This 18% difference from the B point value is not too surprising; the clustering effect takes place here as well.

C=C-treated silicas. The vinylsilane was added to the silicas after pumping at 170 °C for HiSil and at 100 °C for Cab-o-Sil to remove physically adsorbed water. At room temperature no reaction occurred as indicated by the readily

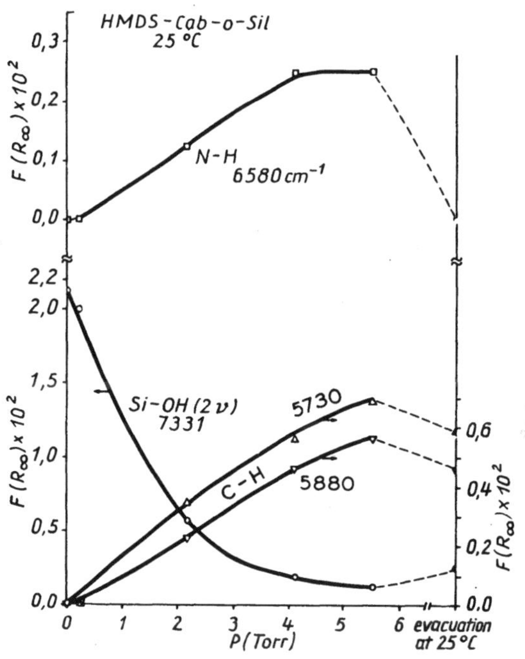

Fig. 10. Band intensities for HMDS adsorption on Cab-o-Sil from fig. 7: ○ free OH band at 7331 cm⁻¹; △ and ▽ C—H bands at 5730 and 5880 cm⁻¹; □ N—H(NH₃) band at 6580 cm⁻¹. Dashed lines show changes caused by evacuation at room temperature to 10⁻⁶ torr

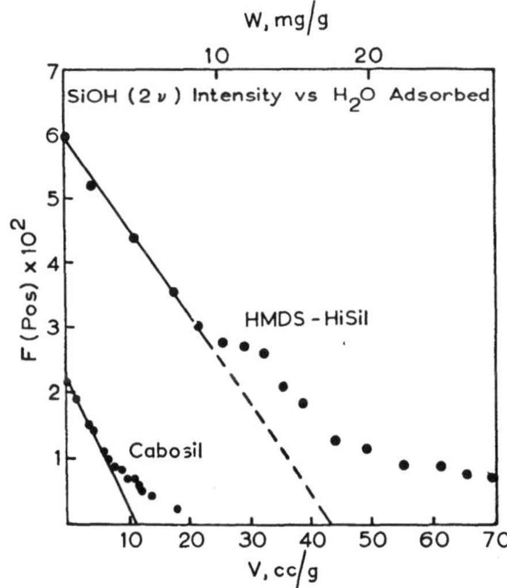

Fig. 11. Band intensities (S-K-M function) for water on 100% HMDS-treated HiSil ○ and on bare Cab-o-Sil. The tailing off is due to clustering of water molecules around those first adsorbed

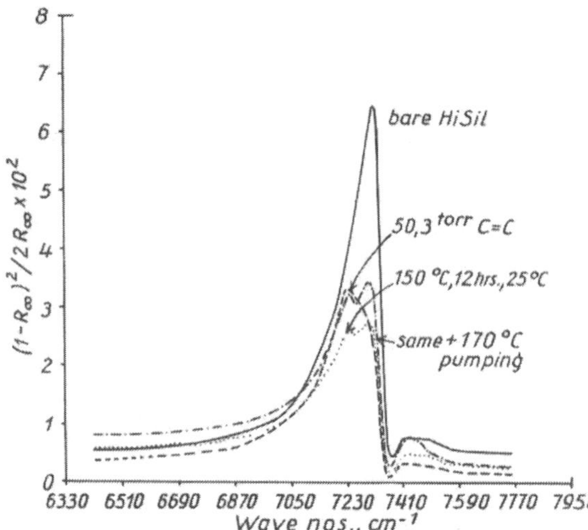

Fig. 12. Spectra showing both the perturbation of the OH absorption peak (2ν) by physisorbed and by chemisorbed vinylsilane

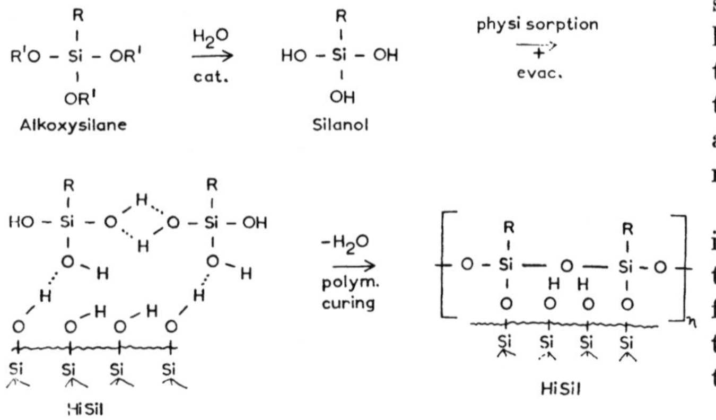

Fig. 13. Schematic diagram for the reaction of alkoxysilanes with a highly hydroxylated silica surfaces, HiSil. A two dimensional polymer is eventually formed upon heat treatment

pumped off C=C leaving the bare silica OH stretch peak.

However, the physically adsorbed C=C perturbed the OH stretch frequency on HiSil to 7230 cm^{-1} at about 50 torr as seen in Fig. 12. After reaction at 170 °C for 12 hours, the residual OH frequency was at 7252 cm^{-1} as perturbed by the reacted ligand (14). The fact that much OH still persists after reaction suggests that not all the Cl's are reacted, as might be expected. The reaction of the C=C compound with the surface OH's is much weaker than that of HMDS.

Cab-o-Sil reacted at 110 °C or 170 °C for 12 hours also showed a small amount of residual OH.

For both silicas, the frequencies at 5900 and 6125 cm^{-1} are assigned to the C—H vibrations. These values are higher than reported for HMDS in accord with the fact that unsaturated groups yield higher frequency C—H vibrations (15).

Water adsorption will be discussed later.

Silanetriol-treated silicas. Three were reacted with the silicas from water solutions. The starting compounds were the alkoxy compounds (table 2) which were hydrolyzed by the addition of several drops of acetic acid to the triols. After 3 hours of reaction with the silicas at room temperature, the solids were centrifuged and then pumped off at 50 °C in a vacuum oven.

The reaction with HiSil is depicted in fig. 13. Here the scheme suggests a one layer addition in which the silanol groups condense to form what might be regarded as a two-dimensional polymer. Hydrogen bonding with the unreacted surface OH's is believed to be responsible for holding the silanetriol to the surface to develop the two-dimensional effect. On the contrary, a three-dimensional network develops on Cab-o-Sil as shown in fig. 14 when there are not too few residual OH's to confine the triol to one layer.

There are several pieces of evidence supporting these conclusions. Firstly, the bands such as those for the C—H were about the same intensity for the silanetriol treated HiSil as for HMDS treated HiSil. A greater intensity occurred for the silanetriol treated Cab-o-Sil. An interesting

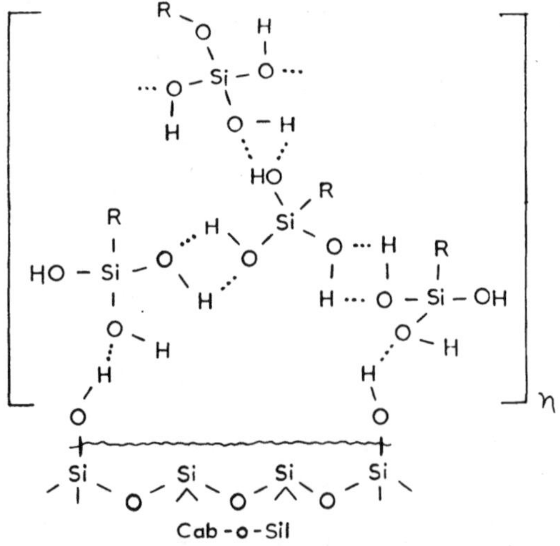

Fig. 14. Schematic showing three-dimensional polymer network that develops when extra OH's are not available, i.e. for Cab-o-Sil, to hold the silanetriol to the surface

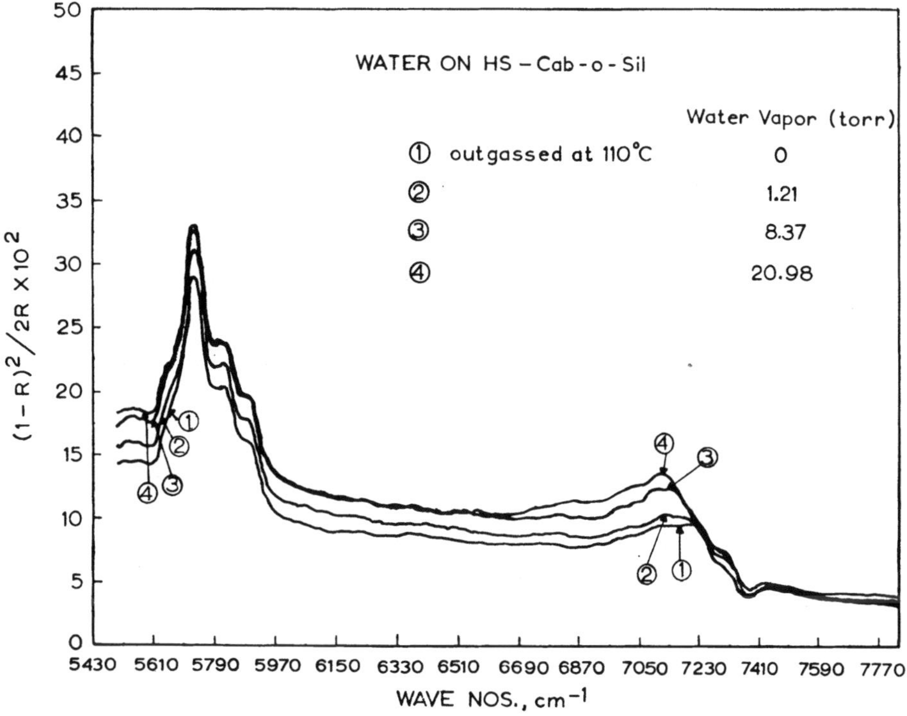

Fig. 15. NIR spectrum for HS treated Cab-o-Sil showing the effect of increased water adsorption

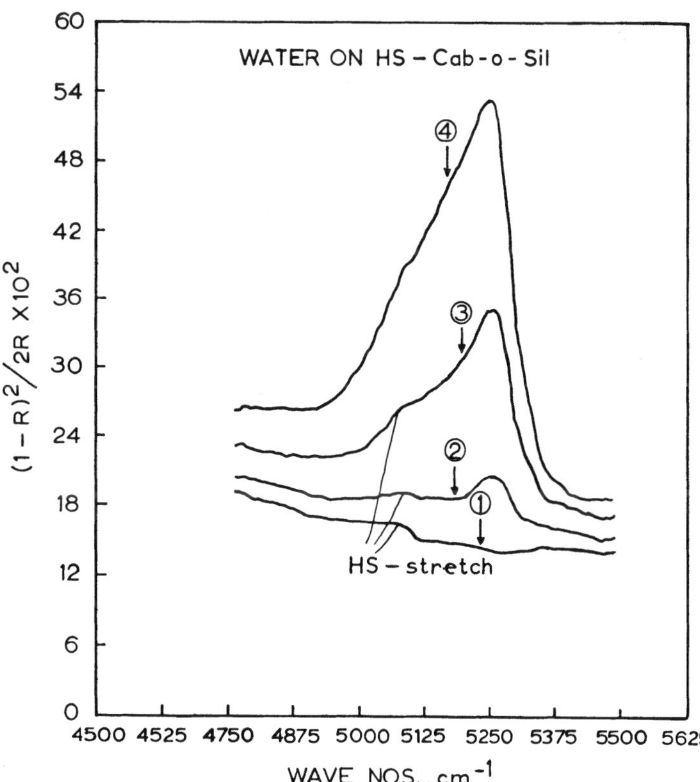

Fig. 16. The water $(\nu + \delta)$ region showing the HS-vibration shoulder which is finally swamped by the combination band of the water

support for this contention is shown for the HS compound in fig. 15, the left-hand portion magnified in fig. 16, both for the Cab-o-Sil treated surface. This band is very weak even for mercaptan compounds and is finally over-whelmed here by the $(\nu + \delta)$ band for water at high coverages. This HS-stretch frequency cannot be detected in the case of the HS-treated HiSil. Thus, the contention that a thicker layer, perhaps 7 or 8 molecules thick as *Lee* has suggested (10), formed from the triols on Cab-o-Sil, whereas only about 1 layer formed on HiSil.

Secondly, this remarkable difference produced on the two substrates was also confirmed by the greater apparent agglomeration of the Cab-o-Sil when treated with the silanetriols. The cross-linking of particles would reduce specific surface area. Measurements showed a reduction of 30—45% for the treated Cab-o-Sils, only 10—15% for the treated HiSils. The conclusion seems inescapable that there is a large difference between the behavior of the sparsely OH populated silica and the abundantly populated silica. It seems plausible that glass fibers might perform better in composites if first stripped of some of their surface OH's.

Table 5 presents the band assigments for the various silane-treated silicas other than those for HMDS (table 4).

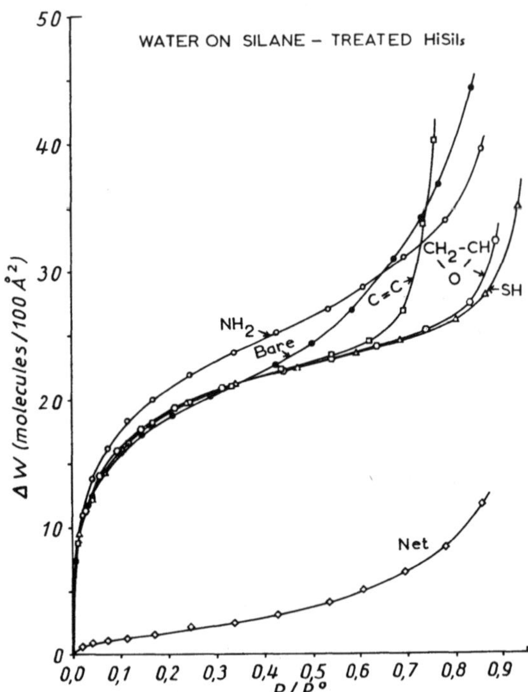

Fig. 17. Water adsorption isotherms for the silane-treated HiSils. Except for the higher adsorption on the NH₂-treated surface, the isotherms are similar in the B-point region. The "net" curve shows the difference in water adsorption between the polar NH₂-treated and the HS-treated surface. (The rise at 0.7 relative pressure for the C=C-treated surface is unexplained.)

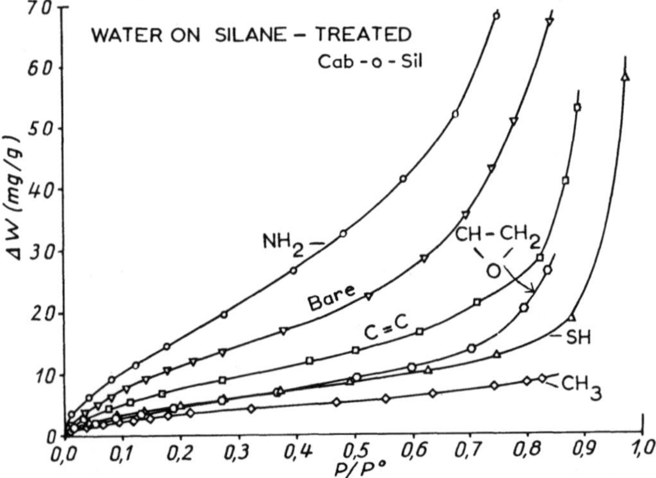

Fig. 18. Water adsorption isotherms for the silane-treated Cab-o-Sils. A distinct order to the polarity of these surfaces now emerges

Table 5. NIR bands: Silane — Treated Silicas

Freq. cm⁻¹	Ligand surface	Assignments — Overtones
4948	NH₂—HiSil NH₂—Cab-o-Sil	NaH₂-bending plus stretch combination band
5079	HS—Cab-o-Sil	H—S stretch
5670	NH₂—HiSil	C—H stretch
5820	NH₂—Cab-o-Sil	
5734	HS—HiSil	C—H stretch
5821, 5886	HS—Cab-o-Sil	
5737, 5858	CH₂—CH—HiSil ＼／ O	C—H stretch
6086	CH₂—CH—Cab-o-Sil ＼／ O	
5900, 5959	C=C—HiSil	C—H stretch
6125	C=C—Cab-o-Sil	
6540	NH₂—HiSil NH₂—Cab-o-Sil	N—H stretch
7225	NH₂—Cab-o-Sil	O—H stretch —H bonded
7252	C=C—HiSil	O—H stretch — free OH's
7270	NH₂—HiSil	same
7283	HS—HiSil	same
7300	C=C—HiSil	same
7331	C=C—Cab-o-Sil	same

Water isotherms for the silane-treated HiSils are presented in fig. 17. These have been corrected to areas determined for each. They show the same B-point except for the NH_2-treated HiSil which shows an enhanced specificity for water. The curve marked "net" is the difference from the HS-treated and the NH_2-treated HiSil curves. The R-groups restricted the growth of the water multilayers except for the $C=C$ case where the sharp rise at 0.7 relative pressure remains unexplained.

The silane-treated Cab-o-Sils yield unique water isotherms, fig. 18, for each ligand. Except for the $HMDS(CH_3)$ and the $C=C$ treated Cab-o-Sil, the layers are thick so that there is no point in putting the results on a unit area basis. The NH_2-treated sample required three to five times the equilibrium time of the others indicating diffusion into the thick layer. Rating the hydrophobicity of the ligands is distinct because the surface hydroxyls have essentially disappeared, but the presence of multilayers in the case of the silanetriols makes a 1:1 analysis impossible. It is interesting that the $C=C$ ligand showed considerably higher water attraction than the CH_3 ligand.

A number of subtle differences in the water adsorption NIR spectra have also been found, differences which will not be discussed here.

Acknowledgement

This work was supported by the Industrial Liaison Program of the Center for Surface and Coatings Research, Lehigh University.

Summary

Both a fully hydroxylated silica (HiSil 233) and a partially hydroxylated silica (Cab-o-Sil M-5) were treated with organosilanes and a silazane. Water adsorption isotherms were then monitored on the resulting products. Reflectance IR spectroscopic measurements were used to follow the reactions of the surface hydroxyls with the silanes and to examine the subsequent water adsorption. The main objective was to ascertain the water susceptibilities of the different functional groups and the availability to water of any residual hydroxyls.

The silazane and silanes employed were: hexamethyl-disilazane, vinyl-trichlorosilane and the alkoxysilanes: γ-glycidoxypropyltrimethoxysilane, 3-mercaptopropyl-trimethyoxysilane and aminopropyltriethoxy-silane. The latter three were reacted from the liquid phase, first hydrolyzing to the silanetriols, with additional complications from condensation reactions. The vinyl compound and the disilazane were reacted from the vapor phase.

Sharp difference between the results for the fully hydroxylated and the partially (25%) hydroxylated silica were found. The silanetriols upon heat treatment formed a multilayer polymer network on the latter and only a two-dimensional network on the former apparently due to hydrogen bonding to residual hydroxyls. The reacted silanes formed umbrellas over reacted hydroxyls on the former which were nevertheless available to water molecules. With the latter, however, a much more hydrophobic surface was produced. A number of details were also provided by the reflectance IR measurements. Finally, the following order of susceptibility to water molecules was established, particularly for the treated Cab-o-Sils:

$$-OH(HiSil) > -OH(Cab\text{-}o\text{-}Sil) \geqq -NH_2 \gg -C=C$$
$$> -SH \geqq CH_2-CH- > -CH_3 .$$
$$\diagdown\diagup$$
$$O$$

References

1) *Bassett, D. R., E. A. Boucher*, and *A. C. Zettlemoyer*, J. Colloid Interface Sci. **27**, 649 (1968).
2) *Bassett, D. R., E. A. Boucher*, and *A. C. Zettlemoyer*, J. Colloid Interface Sci. **34**, 436 (1970).
3) *Zettlemoyer, A. C.*, J. Colloid Interface Sci. **28**, 4 (1968).
4) *Klier, K., J. H. Shen*, and *A. C. Zettlemoyer*, J. Phys. Chem. **77**, 1458 (1973).
5) *Stark, F. O., O. K. Johannson, G. E. Vogel, R. G. Chaffee*, and *R. M. Lacefield*, J. Phys. Chem. **72**, 2750−2754 (1968).
6) *Hertl, W.* and *M. L. Hair*, J. Phys. Chem. **75**, 181 (1971).
7 a) *Lygin* and *A. V. Kiselev* (Moscow), Kolloid. Zhur. **23**, 250−253 (1961).
7 b) *Davydov, V. Y., A. V. Kiselev*, and *L. T. Zhuravlev*, Trans. Faraday Soc. **60**, 2254 (1964).
8) *Snyder* and *Ward* (Union Oil), J. Phys. Chem. **70**, 3941−3952 (1966).
9) *Hair, M. L.* and *W. Hertl* (Corning), J. Phys. Chem. **73**, 2372−2378 (1969).
10) *Lee, L. H.* (Dow), J. Colloid Interface Sci. **27**, 751 − 760 (1968).
11) *Kharilonov, N. P., N. E. Glushkova*, and *A. S. Zhukova*, Iz. Akad. Nauk SSSR, Neorgan. Mat. **6**, 59 − 62 (1970).
12) *Baker, F. S.* and *K. S. W. Sing*, Paper R7, 49th National Colloid Symposium, Clarkson College, June 16−18, 1975 (to be published, J. Colloid Interface Sci.).
13) *Hair, M. L.*, Infrared Spectroscopy in Surface Chemistry (New York 1967).
14) *Wall, T. T.* and *D. F. Hornig*, J. Chem. Phys. **43**, 2079 (1965).
15) *Bellamy, L. J.*, Infrared Spectra of Complex Molecules (London 1958).

Authors' address:

H. H. Hsing and *A. C. Zettlemoyer*
Center for Surface and Coating Research,
Lehigh University,
Bethlehem, Pennsylvania 18015 (USA)

Progr. Colloid & Polymer Sci. **61**, 64–70 (1976)
© 1976 by Dr. Dietrich Steinkopff Verlag GmbH & Co. KG, Darmstadt
ISSN 0340-255 X

Plenary lecture of the IUPAC-Conference on Colloid and Surface Science in
Budapest, September 15–20, 1975

Technical University of Denmark, Lyngby (Denmark)
and Free University of Brussels, Brüssel (Belgium)

Surface chemical and hydrodynamic stability

T. S. Sørensen, M. Hennenberg, A. Steinchen-Sanfeld, and *A. Sanfeld*

With 19 figures

(Received December 9, 1975)

Introduction

In the present paper, we have analyzed the chemical and hydrodynamical stability of an interface when transfer of matter across the surface and chemical reactions occur between macromolecules spread at the surface.

The purpose of this work is to show that interfacial reactions proceeding far from equilibrium, are able to induce surface deformations. The treatment was done for plane and spherical interfaces, but the results we will give here will deal only with the plane interface. For the spherical interface, we have to improve the model to take into account some effects due to the change of the metrics that could be important.

What is the role of macromolecular reactions and transfer of matter in the deformation process of a surface?

As we know from the physical chemistry of interfaces (1, 2), the concentration of molecules of large molecular weight in the surface, greatly affects the surface tension. Processes such as transfer of matter from the bulk phases to the surface or chemical reactions of conformational change of macromolecules in the surface will thus be responsible for changes in the surface tension. This last quantity plays a fundamental role in the hydrodynamics of the neighbourhood of the interface: a local change of surface tension is able to induce motion in the surface (Marangoni effect) as well as a local change of the radius of curvature (Laplace effect).

The deformation of the surface is directly connected with the hydrodynamics of the volume elements near the interface. Indeed, a deformation of the interface implies the inset of motion in the fluids i.e. a velocity of the neighbouring volume elements that "push" on both sides of the surface. The surface tension appears thus as having a chemical — as a hydrodynamical character. The analysis of the coupling between transfer and reaction processes and the hydrodynamic motion, through the surface tension, will permit us to obtain the conditions for which the system will become unstable and get deformations.

What are the implications of this phenomenon in biology?

The cell membrane is now considered as being a more or less fluid medium in which proteins can move through lipidic entities (3). Transfer of matter and membrane reaction could affect the surface tension of the cell and could be responsible for the primary process of cell deformation in the phagocytosis and in the cell mobility (4).

The primary deformation process could allow the contraction of the microfilaments of the endoplasmic reticulum that could have a positive feedback on the primary process. The study of the over-all process of cell deformation is not the purpose of the present paper but we now proceed with a model that could also account for the microfilaments contraction.

1. Chemical kinetics and fluxes of matter in the surface

Let us consider a plane layer of long chain macromolecules (like for example in a cell membrane — glycoproteins and phospholipids) separating two immiscible fluid phases.

In accordance with the rheological conceptions of *Singer-Nicolson* (5) and *Capaldi* (6) generally admitted for the membranes, we will consider this layer as a relatively fluid system.

The main physical properties of this layer we will consider here are its surface tension σ and adsorption Γ and its surface viscosity $(\varkappa + \varepsilon)$.

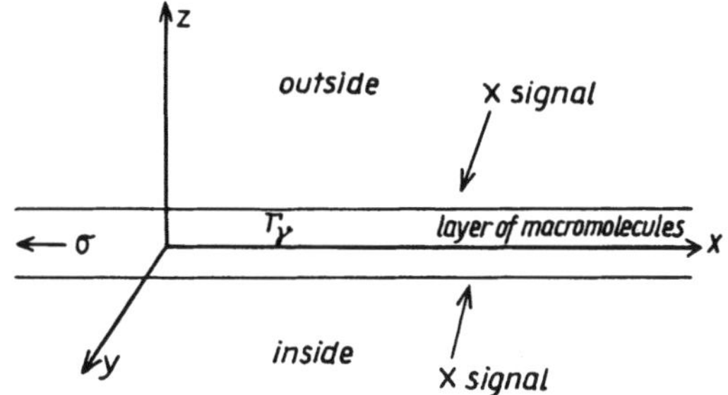

Fig. 1. See text

Let us now consider that a chemical signal coming from outside (e) or from inside (i) reaches the layer and induces a chemical reaction into this surface phase. The excess concentrations Γ_γ of the substances γ will change as a result of the fluxes of matter J_γ and a source term due to the chemical reaction $F_\gamma(\Gamma_\beta)$. Furthermore there will be two-dimensional diffusion. (Fick law)

$$D_\gamma^s \left(\frac{\partial^2}{\partial x^2} + \frac{\partial^2}{\partial y^2} \right) \Gamma_\gamma$$

and convection

$$\left(\frac{\partial}{\partial x} v_x \Gamma_\gamma + \frac{\partial}{\partial y} v_y \Gamma_\gamma \right)$$

can occur in the surface.

The balance of each component γ in the surface is thus easy to derive:

$$\frac{\partial}{\partial t} \Gamma_\gamma = \underset{\text{reactions}}{F_\gamma \{\Gamma_\beta\}} + \underset{\text{fluxes}}{J_\gamma} + D_\gamma^s \underset{\text{surface diffusion}}{\left(\frac{\partial^2}{\partial x^2} + \frac{\partial^2}{\partial y^2} \right)} \Gamma_\gamma$$
$$- \underset{\text{surface convection}}{\frac{\partial}{\partial x}(v_x \Gamma_\gamma)} - \frac{\partial}{\partial y}(v_y \Gamma_\gamma) - \underset{\text{change of metrics}}{\Gamma_\gamma \frac{\dot a}{2a}}$$

Fig. 2. See text

The last term in fig. 2 is not important for a plane layer for very small deformations as we shall study here. It only takes importance for curved surfaces.

The surface tension of the layer is related to its chemical composition by a state equation

$$\underset{\uparrow \text{ surface tension without layer}}{\overset{\downarrow \text{ surface tension}}{\sigma = \sigma_0 - \left(\sum \alpha_\gamma \Gamma_\gamma + \text{virial} \right)}}$$
$$\delta_0 = - \sum \alpha_\gamma \, \delta \Gamma_\gamma$$

Fig. 3. See text

2. Hydrodynamic processes

Now, how does the surface deform itself? Let us suppose that the chemical reaction induces a local change in the surface tension. At that point, the radius of curvature will change and the volume element adjacent to the surface will move i.e. convection will start in the bulk phases near the interface.

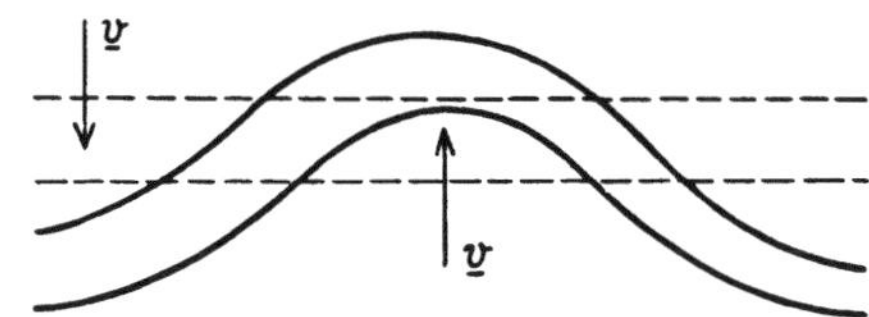

Fig. 4. See text

This motion in the bulk phases is described by the well known Navier-Stokes equation

$$\underset{\text{density}}{\overset{\downarrow}{\varrho}} \frac{\mathrm{d}}{\mathrm{d}t} v = - \underset{\text{hydrost. pressure}}{\overset{\downarrow}{\text{grad } p}} + \underset{\text{external force}}{\overset{\downarrow}{F \varrho}} + \underset{\text{viscous tensor}}{\overset{\downarrow}{\text{Div } \boldsymbol{\Pi}}}$$

or for Newtonian incompressible fluids

$$\varrho \frac{\mathrm{d}}{\mathrm{d}t} v = - \text{grad } p + F \varrho + \underset{\text{Laplace operator}}{\overset{\overset{\text{viscosity coefficient}}{\downarrow}}{\mu \, \Delta v}}$$

Fig. 5. See text

The two Navier Stokes equations have to be connected through boundary conditions at the surface.

1. There must be no cavitation, i.e. the velocity and its space first derivative has to be continuous.
2. The difference of hydrostatic pressure across the surface is balanced by the surface tension (7) (Laplace law).
3. The difference of the lateral stresses on both sides of the surface is balanced by the

$$v_z^{\mathrm{e}}\big|_{z=0} = v_z^{\mathrm{i}}\big|_{z=0} = v_z^{\mathrm{s}}$$

$$\left.\left(\frac{\partial}{\partial z}\,v_z\right)^{\mathrm{e}}\right|_{z=0} = \left.\left(\frac{\partial}{\partial z}\,v_z\right)^{l}\right|_{z=0} = \left(\frac{\partial v_z}{\partial z}\right)^{s} \Bigg\}\ \text{no cavitation}$$

$$p^{\mathrm{i}} - p^{\mathrm{e}} = \sigma\left(\frac{1}{R_1} + \frac{1}{R_2}\right) + \overset{\text{viscosity terms bulk}}{\Pi_{zz}^{\mathrm{e}}} - \Pi_{zz}^{\mathrm{i}}$$

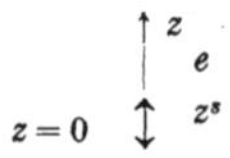

$$+\,F_z\,\Gamma_T + \cancel{s}\,F_z(\varrho^{\mathrm{e}} - \varrho^{\mathrm{i}}) - \underbrace{\Gamma_T\,\frac{\mathrm{d}v_z^{s}}{\mathrm{d}t}}\quad \text{(Laplace condition)}$$

$$\underbrace{}_{\text{external force}} \qquad \underbrace{\phantom{\Gamma_T\,\frac{\mathrm{d}v_z^{s}}{\mathrm{d}t}}}_{\text{acceleration}}$$

shear stress in bulk

$$\Pi_{xz}^{\mathrm{e}} - \Pi_{xz}^{\mathrm{i}} = \frac{\partial\sigma}{\partial x} + (\varkappa + \varepsilon)\left(\frac{\partial^2 v_x}{\partial x^2} + \frac{\partial^2 v_x}{\partial y^2}\right) - \Gamma\,\frac{\mathrm{d}v_x}{\mathrm{d}t}$$

$$\Pi_{yz}^{\mathrm{e}} - \Pi_{yz}^{\mathrm{i}} = \frac{\partial\sigma}{\partial y} + \overset{\text{surface viscosity}}{(\varkappa + \varepsilon)}\left(\frac{\partial^2 v_y}{\partial x^2} + \frac{\partial^2 v_y}{\partial y^2}\right) - \Gamma\,\frac{\mathrm{d}v_y}{\mathrm{d}t}\quad \text{(Levich-Aris condition)}$$

$$\frac{\partial}{\partial t}\,\Gamma_\gamma = F_\gamma\{\Gamma_\beta\} + F_\gamma + D_\gamma^{s}\left(\frac{\partial^2}{\partial x^2} + \frac{\partial^2}{\partial y^2}\right)\Gamma_\gamma - \frac{\partial}{\partial x}(v_x\,\Gamma_\gamma) - \frac{\partial}{\partial y}(v_y\,\Gamma_\gamma)$$

$$\text{(chemical kinetics)}$$

Fig. 6. See text

gradient of surface tension and the surface viscosities (Levich-Aris condition (8, 9)).

4. The equations of the chemical kinetics and diffusion-convection mass balances.

3. Stability analysis

A linearized perturbation analysis of the equations of hydrodynamics with their boundary conditions and of the chemical kinetics in the surface, permits to describe the evolution of the perturbations of velocity and of adsorption around a steady state of reference corresponding to the mechanical equilibrium.

If these perturbations are growing in time the system is unstable and the surface will deform. The solutions of these perturbation equations are of the following type (Fourier development restricted to one normal mode).

The real part of ω characterizes the stability: if $\operatorname{Re}\omega > 0$ the system is unstable.

The imaginary part is responsible for temporal oscillations. We have performed this type of analysis for various kinds of surface reactions:

1. for an autocatalytic reaction (10);

$$\delta v_z = \tilde{v}_z(z)\,\mathrm{e}^{\omega t + \mathrm{i}k_x x + \mathrm{i}k_y y}$$

$$\delta\Gamma_\gamma = \widetilde{\Gamma}_\gamma\,\mathrm{e}^{\omega t + \mathrm{i}k_x x + \mathrm{i}k_y y}$$

with $\tilde{v}_z(z)$ and $\widetilde{\Gamma}_\gamma$, the amplitudes of the perturbations

ω the frequency of the perturbation

$$k_x^2 + k_y^2 = k^2 \quad \text{with} \quad k = \frac{2\pi}{\lambda}\text{, the wave}$$

number

Fig. 7. See text

2. for a transconformation electrochemical reaction (4, 11) (allosteric enzyme type scheme);
3. for any type of reaction scheme (12).

Let us now analyse the results obtained for the general kinetic equation of fig. 10. The perturbation analysis gives us a secular equation that relates the frequency of the perturbation ω to the wavenumber k and the parameters of the system (viscosities, densities, surface tension, surface concentrations in the reference state).

$$A \xrightarrow{k_1} X \qquad 2X + Y \xrightarrow{k_2} 3X$$

$$B + X \xrightarrow{k_3} Y + D \qquad X \xrightarrow{k_4} E$$

$$\frac{\partial \Gamma_x}{\partial t} = k_1 \Gamma_A + k_2 \Gamma_x^2 \Gamma_y - k_3 \Gamma_B \Gamma_x$$

$$\frac{\partial \Gamma_y}{\partial t} = k_3 \Gamma_B \Gamma_x - k_2 \Gamma_x^2 \Gamma_y$$

Fig. 8. Autocatalytic trimolecular scheme

$$\bigcirc \underset{k_c'}{\overset{k_c}{\rightleftharpoons}} \square \quad \text{cooperative transconformation}$$

$$\left.\begin{array}{l} \square + A \underset{k_d^i}{\overset{k_a^i}{\rightleftharpoons}} \boxdot \\[2mm] \square + B \underset{k_d^e}{\overset{k_a^e}{\rightleftharpoons}} \boxdot \end{array}\right\} \text{fixation}$$

$$\left.\begin{array}{l} A^i \underset{k_{-p}^i}{\overset{k_r^i}{\rightleftharpoons}} A \\[2mm] B^e \underset{k_{-p}^a}{\overset{k_r^e}{\rightleftharpoons}} B \end{array}\right\} \begin{array}{l} \text{diffusion-migration from} \\ \text{the bulk} \end{array}$$

$$\frac{\partial \Gamma_{\square + \boxdot}}{\partial t} = k_c [\Gamma_T - \Gamma_{\square + \boxdot} - e^{-A/RT} e^{\eta \, \Gamma_{\boxdot} + \square / \Gamma_T} (\Gamma_{\square + \boxdot} - \Gamma_{\boxdot})]$$

Fig. 9. Allosteric type scheme

$$\frac{\partial \Gamma_\gamma}{\partial t} = \sum_r \sum_\beta k_r \Gamma_\beta^n \Gamma_\gamma^m \quad \begin{array}{l}\text{general kinetic}\\ \text{equation}\end{array}$$

for the perturbation

$$\frac{\partial \, \delta \Gamma}{\partial t} = \sum_\beta c_{\gamma\beta} \, \delta \Gamma_\beta = F_\gamma \{\delta \Gamma_\beta\} \quad \text{kinetic matrix}$$

where $c_{\gamma\beta}$ include the kinetic constants and the concentrations in the reference state with the appropriate exponent.

Fig. 10. See text

The solutions of this equation gives us the explicit form of the function

$$\omega = \omega(k, \mu, \varepsilon + \varkappa, \Gamma_\gamma^0 \cdots \sigma^0) .$$

We then can calculate the value of ω for different wavenumbers and for different values of the parameters of the system, and see if the system is oscillating or not and if it is stable or not.

In the general case, for one fluctuating species at the surface, the secular equation can be solved manually, for certain assumptions on the wavelength.

Indeed

For some small wavelength of particular values

$$\frac{\omega \varrho}{\mu} \ll k^2 \le \frac{\omega^2 \left(\Gamma_T + \dfrac{\varrho^i + \varrho^e}{2k}\right)}{(\varepsilon + \varkappa) \, |(k^2 D_s - c)|}$$

where Γ_T is the total surface concentration $= \sum \Gamma_\gamma$

$$\left\{\omega^2 \left[\Gamma_T + \frac{3}{2} \frac{\varrho^i + \varrho^e}{k}\right] + 2k\,\omega(\mu^i + \mu^e) + \sigma k^2 + g(\varrho^i - \varrho^e)\right\}$$
$$\times \left\{\omega^2\left(\Gamma_T + \frac{\varrho^i + \varrho^e}{2k}\right) + \omega\left[\left(\Gamma_T + \frac{\varrho^i + \varrho^e}{2k}\right)(k^2 D_s - c) + k^2(\varepsilon + \varkappa) + 2k(\mu^i + \mu^e)\right]\right.$$
$$\left. + (k^2 D_s - c)[k^2(\varepsilon + \varkappa) + 2k(\mu^i + \mu^e)] + \alpha k^2 \Gamma\right\} = 0$$

Fig. 11. See text

The vanishing of the first factor in this equation gives the classical solutions of the Rayleigh-Taylor problem for the small wavelengths: i.e. that the state of marginal stability ($\omega = 0$) is reached for $\sigma k^2 = g(\varrho^e - \varrho^i)$ if $\varrho^e > \varrho^i$. It means that if the density of the upper fluid is larger than the density of the inferior fluid, the system will become unstable (the surface will break down) for all the perturbations having a wavenumber smaller than the critical value

$$k_c = \left(\frac{g(\varrho^e - \varrho^i)}{\sigma}\right)^{1/2} .$$

The vanishing of the second factor of fig. 11 leads to another marginal state ($\omega_r = \omega_i = 0$) with exchange of stability:

$$(k^2 D_s - c)[k^2(\varepsilon + \varkappa) + 2k(\mu^i + \mu^e)] + \alpha k^2 \Gamma = 0$$

or

$$(c - k^2 D_s) = \frac{\alpha k \Gamma}{k(\varepsilon + \varkappa) + 2(\mu^i + \mu^e)}$$

this equality holds only if

$$c > k^2 D_s$$

Fig. 12. See text

Here we see the natural compensation of the chemical effect by the stabilizing effects of the surface activity (if α is great the substance is very surface active) and the surface and bulk viscosities.

A necessary but not sufficient condition for the system to become unstable is thus $c > k^2 D_s$ (chemical instability), because of the damping effect of the viscosities and of the tensioactivity. A marginal state with oscillations ($\omega_r = 0$, $\omega_i \neq 0$) can be obtained when the coefficient of ω in the second factor of fig. 11 is equal to zero i.e.:

if $c > k^2 D_s$ and

$$(k^2 D_s - c) = - \frac{k^2(\varepsilon + \varkappa) + 2k(\mu^i + \mu^e)}{\Gamma_T + \dfrac{\varrho^i + \varrho^e}{2k}}$$

we have

$$\omega^2 - \left[\frac{k^2(\varepsilon + \varkappa) + 2k(\mu^i + \mu^e)}{\Gamma_T + \dfrac{\varrho^i + \varrho^e}{2k}} \right]^2 + \frac{\alpha k^2 \Gamma}{\Gamma_T + \dfrac{\varrho^i + \varrho^e}{2k}} = 0$$

if

$$\alpha k^2 \Gamma > \frac{[k^2(\varepsilon + \varkappa) + 2k(\mu^i + \mu^e)]^2}{\Gamma_T + \dfrac{\varrho^i + \varrho^e}{2k}}$$

$$= (c - k^2 D_s) \cdot (k^2(\varepsilon + \varkappa) + 2k(\mu^i + \mu^e))$$

the solutions are pure imaginary numbers
$\Rightarrow$ overstability = oscillations

Fig. 13. See text

It means that the perturbations of v and of Γ are periodic functions of time.

According to the thresholds of instability given in figs. 12 and 13, the system can be unstable without oscillations ($\omega_r > 0$ and $\omega_i = 0$) in a range of wavelengths and for peculiar values of the parameters: for example

For

$$\mu^i = \mu^e = 10^{-1} \text{ poise} \quad \varrho^i = \varrho^e = 1 \text{ g/cm}^3$$
$$\varepsilon + \varkappa = 1 \quad \text{surface poise}$$
$$\Gamma_T = 10^{-4} \text{ g/cm}^2 \qquad \Gamma = 10^{-6} \text{ g/cm}^2$$
$$D_s = 10^{-6} \text{ cm}^2/\text{sec}$$
$$c = 10^5 \text{ sec}^{-1} \qquad \alpha = 10^8 \text{ cm}^2/\text{sec}^2$$

instabilities are obtained for

$$4.5 \times 10^{-6} \text{ cm} > \lambda > 3.2 \times 10^{-6} \text{ cm}$$

Fig. 14. See text

Now, for another range of small wavelengths, the secular equation can be reduced to a most simple form; that is:

if $\dfrac{\omega \varrho}{\mu} \ll k^2$

but $k^2 \geq \dfrac{\omega^2 \left(\Gamma_T + \dfrac{\varrho^i + \varrho^e}{2k} \right)}{(\varepsilon + \varkappa) \, | (k^2 D_s - c) |}$

then we have

$$\left\{ \omega^2 \left[\Gamma_T + \frac{3}{2} \frac{\varrho^i + \varrho^e}{k} \right] + 2k \omega (\mu^i + \mu^e) + \sigma^0 k^2 + g(\varrho^i - \varrho^e) \right\}$$

$$\times \left\{ \omega \left[\left(\Gamma_T + \frac{\varrho^i + \varrho^e}{2k} \right) (k^2 D_s - c) + k^2(\varepsilon + \varkappa) + 2(\mu^i + \mu^e) k \right] \right.$$

$$+ (k^2 D_s - c)[k^2(\varepsilon + \varkappa) + 2(\mu^i + \mu^e) k]$$

$$\left. + \alpha k^2 \Gamma \right\} = 0$$

Fig. 15. See text

The vanishing of the second factor of this equation leads to a real solution (no oscillation) given by fig. 16.

$$\omega = -\ \frac{(k^2 D_s - c)\,[k^2(\varepsilon + \varkappa) + 2\,(\mu^{\mathrm{i}} + \mu^{\mathrm{e}})\,k] + \alpha\,\Gamma\,k^2}{\left(\Gamma_T + \dfrac{\varrho^{\mathrm{i}} + \varrho^{\mathrm{e}}}{2\,k}\right)(k^2 D_s - c) + k^2(\varepsilon + \varkappa) + 2\,(\mu^{\mathrm{i}} + \mu^{\mathrm{e}})\,k}$$

for example: if

$\varrho^{\mathrm{i}} = \varrho^{\mathrm{e}} = 1\ \mathrm{g/cm^3}, \quad \mu^{\mathrm{i}} = \mu^{\mathrm{e}} = 10^{-2}\ \mathrm{poise} \quad \varepsilon + \varkappa = 1\ \mathrm{surface\ poise}$
$\Gamma_T = 10^{-6}\ \mathrm{g/cm^2}, \quad \Gamma = 10^{-7}\ \mathrm{g/cm^2} \quad D_s = 10^{-5}\ \mathrm{cm^2/sec}$
$c = 10^5\ \mathrm{sec^{-1}}, \quad \alpha = 10^8\ \mathrm{cm^2/sec^2}$

the system is unstable for $10^{-4}\ \mathrm{cm} > \lambda > 10^{-5}\ \mathrm{cm}$

Fig. 16. See text

For example for an allosteric type scheme like in fig. 9 the value of the kinetic parameter c is given by

$$c = -\ k_c \left[1 - \eta \left(1 - \frac{\Gamma}{\Gamma_T} \right) + \mathrm{e}^{-A/RT}\,\mathrm{e}^{-\eta\,\Gamma/\Gamma_T} \right]$$

kinetic constant   cooperativity parameter

for $\quad k_c = \dfrac{10^5}{2 \cdot 3}\ \mathrm{sec^{-1}} \quad \eta = 5 \quad \Gamma = 10^{-7}\ \mathrm{g/cm^2}$

$\Gamma_T = 10^{-6}\ \mathrm{g/cm^2} \quad \mathrm{e}^{-A/RT} = 2 \Rightarrow c = 10^5\ \mathrm{sec^{-1}} \quad \mathrm{if} \quad D_s = 10^{-5}\ \mathrm{cm^2/sec}$
$\alpha = 10^8\ \mathrm{cm^2/sec^2} \quad \varepsilon + \varkappa = 1\ \mathrm{surface\ poise} \quad \mu^{\mathrm{i}} = \mu^{\mathrm{e}} = 10^{-2}\ \mathrm{poise}$
$\varrho^{\mathrm{i}} = \varrho^{\mathrm{e}} = 1\ \mathrm{g/cm^3}$

The system is unstable without oscillations for $10^{-4}\ \mathrm{cm} > \lambda > 10^{-5}\ \mathrm{cm}$

Fig. 17. See text

Another case is, if the bulk phases have a very low viscosity $\mu^{\mathrm{i}} = \mu^{\mathrm{e}} = 0$ the secular equation reads

$$\omega^2 = \frac{g\,(\varrho^{\mathrm{i}} - \varrho^{\mathrm{e}}) - \sigma\,k^2}{\Gamma_T + \dfrac{\varrho^{\mathrm{i}} + \varrho^{\mathrm{e}}}{k}}.$$

Non-viscous bulk phases

$$\left\{ \omega^2 \left[\Gamma_T + \frac{\varrho^{\mathrm{i}} + \varrho^{\mathrm{e}}}{k} \right] + \sigma\,k^2 + g\,(\varrho^{\mathrm{i}} - \varrho^{\mathrm{e}}) \right\}$$
$$\times\ \{ \omega^2 \Gamma_T + \omega\,[\Gamma_T (k^2 D_s - c) + k^2(\varepsilon + \varkappa)]$$
$$+\ k^2(\varepsilon + \varkappa)(k^2 D_s - c) + \alpha\,k^2 \Gamma \} = 0$$

Fig. 18. See text

The vanishing of the first factor gives the solution for the Rayleigh-Taylor problem in the non viscous case, i.e.

The presence of the surface acceleration term increases somewhat the value of the wavelength corresponding to the maximum of instability and also has a weak contribution to the enhancement of the instability.

The vanishing of the second term leads to the relation:

$$\omega^2 \Gamma_T + \omega \left[\Gamma_T (k^2 D_s - c) + k^2 (\varepsilon + \varkappa) \right]$$
$$+ k^2 (\varepsilon + \varkappa)(k^2 D_s - c) + \alpha k^2 \Gamma = 0$$

$$\omega = 0 \quad \text{for} \quad k^2 D_s - c = \frac{-\alpha \Gamma}{\varepsilon + \varkappa}$$

$$\omega \text{ is imaginary for } c > k^2 D_s$$

$$\text{and} \quad |k^2 D_s - c| (\varepsilon + \varkappa) > \alpha \Gamma$$

$$\text{with} \quad |k^2 D_s - c| = \frac{k^2 (\varepsilon + \varkappa)}{\Gamma_T}$$

Fig. 19. See text

Here also, a necessary condition for the system to be unstable is $c > k^2 D_s$ i.e. the chemical reaction is unstable in itself. But it is not a sufficient condition because the system has to overcome the stabilizing effects of the surface tension and of the surface viscosity.

4. Conclusion

The method presented here permits us, having a precise mechanism for the chemical reaction, to obtain the conditions under which the system will become unstable, i.e. for which the surface will be deformed. The reaction in itself has to be unstable to obtain a deformation of the surface but it is also necessary to overcome stabilizing effects such as the viscosities of the bulk phases, the viscosity of the surface, the surface tension effects and the surface diffusion.

Zusammenfassung

Die Stabilität mit Rücksicht auf die Deformation einer ebenen Grenzfläche mit grenzflächenaktiven Stoffen oder einer lipiden Doppelschicht mit grenzflächenaktiven Proteinen unter Einfluß von Veränderung der Konformation, Ionenadsorption oder anderen chemischen Reaktionen, ist mit Hilfe der linearen, hydrodynamischen Stabilitätstheorie untersucht.

Im Falle von nur einem fluktuierenden grenzflächenaktiven Stoff muß der entsprechende chemisch-kineti-sche Matrix-Komponent positiv sein (autocatalytische oder cooperative Reaktion), damit das hydrodynamisch-chemische System instabil ist, aber Stufen der Grenzflächendiffusion und Viskosität (gewöhnliche und Grenzfläche) müssen auch überschritten werden.

Summary

The deformational stability of a plane interface with surfactants or of a bilipid layer with surfactive proteins subject to conformation changes, ion adsorption or other chemical reactions is investigated by means of linear, hydrodynamical stability theory.

In case of only one fluctuating surfactant the corresponding component of the matrix of chemical kinetics has to be positive (autocatalytic or cooperative step) for the hydrodynamico-chemical system to be unstable, but thresholds involving surface diffusion and bulk and surface viscosities have also to be surpassed.

References

1) *Defay, R., I. Prigogine,* and *A. Bellemans-Everett,* Surface tension and adsorption.
2) *Davies, J. T.,* Interfacial Phenomena (New York 1961).
3) *Nicolson, G.,* Ser. Haematol. **6**, 275 (1973).
4) *Sanfeld, A.* and *A. Steinchen-Sanfeld,* Biophys. Chem. **3**, 99 (1975).
5) *Singer, S.* and *G. Nicolson,* Science **175**, 720 (1972).
6) *Capaldi, R.,* Scientific Amer. **230**, 27 (1974).
7) *Chandrasekhar, S.,* Hydrodynamic and Hydromagnetic stability, chap. 10 (Oxford 1961).
8) *Levich, V.,* Physico-Chemical hydrodynamics (Englewoods Cliff, New Jersey 1967).
9) *Aris, R.,* Vector, tensors and the basic equations of fluid mechanics, chap. 10 (Englewood Cliffs, New Jersey 1962).
10) *Steinchen-Sanfeld, A.* and *A. Sanfeld,* Chem. Phys. **1**, 156 (1973).
11) *Deyhimi, F.* and *A. Sanfeld,* C. R. Acad. Sc. Paris **279**, 437 (1974).
12) *Hennenberg, M., T. S. Sorensen, A. Steinchen-Sanfeld,* and *A. Sanfeld,* J. Chim. Phys. **72**, 1202 (1975).

Authors' address:

T. S. Sørensen
Technical University of Denmark
Lyngby (Denmark)

M. Hennenberg, A. Steinchen-Sanfeld, and *A. Sanfeld*
Free University of Brussels
Brussels (Belgium)

Progr. Colloid & Polymer Sci. **61**, 71—79 (1976)
© 1976 by Dr. Dietrich Steinkopff Verlag GmbH & Co. KG, Darmstadt
ISSN 0340-255 X

Plenary lecture of the IUPAC-Conference on Colloid and Surface Science in
Budapest, September 15—20, 1975

*Department of Colloid Chemistry, Moscow State University, Moscow (USSR) and
Institute of Physical Chemistry of the Academy of Sciences of the USSR*

Cohesion of particles in disperse systems

E. E. Shchukin and *E. A. Amelina*

With 5 figures and 4 tables

(Received December 9, 1975)

In the solution of the problem of controlling the stability of disperse systems, particularly of regulation of structure formation processes, the main part belongs to a question of interaction of the particles of the disperse phase — the value of cohesion forces (the strength of contacts). The characteristic determines the specificity of rheological and structural-mechanical properties of disperse systems of various type (diluted and concentrated suspensions, dry powders, disperse porous solids).

The interaction of disperse phase particles is determined by the nature of the particles themselves and the environment, the character of interfacial interaction, conditions of boundary layers formation, etc. Investigations of various aspects of interaction of particles in condensed phases have been the subject of works of scientists belonging to principal colloid chemistry schools of the USSR *(Rehbinder, Derjaguin, Lifshitz)*, Holland *(Verwey, Overbeek)*, Hungary *(Buzágh)*, England *(Bradley)*, GDR *(Sonntag)*, Bulgaria *(Sheludko)* and many others.

The subject of this proceeding is connected with a question, which belongs to this problem and, being characteristical for the whole direction of the Rehbinder school, was elucidated mostly lately. It is a matter of some results of experimental studies of cohesion forces between separate particles and of the influence of surface-active substances on their cohesion.

For investigations of contact interactions of particles in disperse systems a method of direct measurement of cohesion forces in contacts between separate solid particles in a wide scale (from 10^{-3} to 10^2 dyne) was worked out in our laboratory.

This method, described in detail in our published works (1, 2) is based upon use of magneto-electric system (galvanometer) as a strength meter (fig. 1). The idea of the experi-

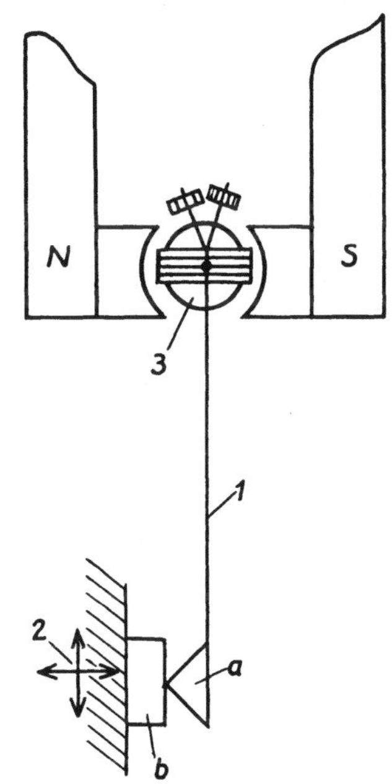

Fig. 1. Scheme of the device

ment was as follows: one particle "a" is rigidly attached to the hand of the galvanometer (1); another — "b" — to the manipulator (2). The particles are brought together with various force (f). The value of this force is given with the intensity of current going through the frame (3). The particles are kept in contact for a definite time. After that a reversed force is applied to them (the reversed current is passed). According to the intensity of current the force required for the separation of particles (destruction of the contact) is registered. This force is the one of cohesion between particles in the contact. We identify it with rupture strength of the contact (p_1, dyne).

Yusupov carried out measurements of cohesion forces in contacts between solid particles of various chemical nature (AgCl, NaCl, naphthalene, anthracene) in air and in liquid environments (3, 4). The compressive force varied from 10^{-3} to 10^1-10^2 dyne. Solid particles 3 mm in size were cleaved out from monocrystals and had natural surface rough as it was. Measurements of cohesion forces between such particles showed, that there takes place wide scatter of cohesion forces in contacts values, the contacts being formed under the same conditions. That is why statistical analysis of the results with consideration of corresponding distributions is necessary. The results of the experiments were presented as histogrammes of the strength of the contacts — the dependence of differential function of distribution $\varrho = \mathrm{d}n/n_0 \, \mathrm{d} \lg p_1$ on strength p_1 logarithm (n and n_0 = current and total number of measurements). Fig. 2 presents the histogramme for AgCl and naphthalene. The axis of abscisses is for contact strength logarithms, that is, corresponding values of strength are

equal to 1000th, 100th, 10th parts of dyne, and so on. The axis of ordinates is for per cent of contacts with the given strength. The scatter of contact strength values is very wide; the dispersion is 2—3 orders. In these experiments such a wide scatter is not due to the experimental technique described. It is determined with the reality of the object and is a result of geometrical and electric unhomogeneity of contacting parts of the surface. As it is seen from the picture, the average strength of contacts for AgCl increases with growth of compressive force between particles. The compressive force f being 0,5 dyne and less, the average strength p_1 is 10^{-3} dyne, and f being 50 dyne — p_1 10 dyne. The strength growth in this case is not smooth, but has a leap. It results in that a second maximum appears on the histograms as the strength reaches some critical value f_c. This leap is connected with conversion of weak contacts (with the strength of 1000th parts of a dyne and less), which corresponds to mere touching of particles, into strong contacts (with the strength of 10th parts of a dyne and more), in which the strength is determined by cohesion forces of short-range interaction. Indeed, the strength of an interatomic bond is the squared charge of electron, divided with the squared interatomic distance, the result being equal approximately to 10^{-4} to a few 10^{-4} dyne. The strength of weak contacts is caused mainly with long-range van-der-Waals forces of attraction. Even if by touching of particles the appearance of valent bonds is possible, cohesion due to them takes place just on the square of one or a few elementary cells. The sharp leaptransition to contacts 0,1 dyne and more strong points out that interaction due to short-range forces involves hundreds and thousands cells. Formation of such phase contacts is the result of plastic deformation in the zone of contact. So the qualitative change of contacts is observed: a point touch transforms and develops into a phase cohesion contact. The contacts being like the latter, continuous transfer from one particle to another takes place *inside one phase*.

It is interesting to compare quantitatively the values of cohesion forces in contacts between separate particles with values of strength of macroscopic samples of porous disperse structures, built of many particles of the same kind. Such a comparison requires us to introduce certain ideas on joint manifestation of cohesion

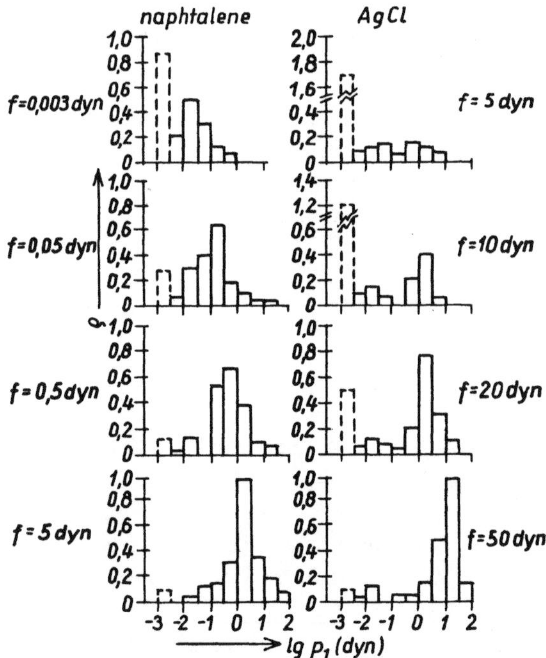

Fig. 2. Histograms of the strength of the contacts, formed between naphthalene and between AgCl-crystals compression of the crystals

forces in contacts. For instance, one can suppose the behaviour of contacts to be additive, i.e. consider, that when the samples are destroyed all the contacts (all the bonds between particles) in the plane of destruction break simultaneously. This concept was experimentally confirmed by *Babak* (5) for model globular structures, obtained by sintering of spheric polystyrene particle. *Yusupov* (6) conducted the comparison of parameters of individual contacts strength distribution between separate crystals — the average value and the dispersion of the distribution — with the value of the strength of disperse structures and its scatter. Direct numerical comparison was held (table 1) between average

Table 1

Force of compression of naphthalene crystals, f dyn	$p_1' = P_c/\chi$ dyn	$\bar{p}_1$ dyn	Temperature of conglomeration of ammonium nitrate crystals, T, °C	p_1' dyn	$\bar{p}_1$ dyn
5	1.0	4.7	100	1.0	0,4
50	12,5	14.6	140	10.0	7.0

strength of inter-particle contact $(\bar{p}_1)$ and approximate value of contacts strength in porous disperse structures p_1': $p_1' = P_c/\chi$, where P_c is the strength of the structure, χ is the number of contacts per unit of cross-section area of the disperse solid; the value of χ was calculated from the data on dispersity of particles and porosity of the samples, the globular model of the porous solid being used (7).

It is shown, that the order of compared values of contacts strength coincide. Moreover, the wide distribution of contacts strength, found as a result of direct measurements of cohesion

forces between separate particles, also correlated with the one for disperse structures.

It was this method, which was used by *Amelina* and *Yusupov* for investigations of cohesion forces between solid AgCl particles in presence of adsorption layers of diphilic surface-active molecules on their surface — cetyl alcohol (CA), decyl alcohol (DA), octadecylamine (ODA), etc. (4, 8). The measurements were conducted in air and in liquid media. In the first case surfactants were spread on the AgCl surface by means of evaporating a drop of known volume of a solution of known concentration of a surface-active substance in heptane. The mean number of monolayers spread was estimated with calculation. Measurements being carried out in liquid media, surfactant layers were spread as a result of equilibrium absorption, when prior to the contact the samples stayed in water or heptane solutions of surfactants; compression and separation of the crystals were carried out in the liquid environment. It was shown that the given liquid media (water and heptane) as they are do not influence the AgCl contacts strength: the histograms were identical. Fig. 3 presents the results of measurements of cohesion forces between AgCl crystals, covered with octadecylamine in air (column II) and in octadecylamine solutions in heptane (column III). Column I is for results of experiments with pure AgCl with no surfactant. It is seen that already in air the surfactant monolayer notably hinders the formation of phase contacts. But the measurements in liquid media demonstrate it especially strikingly. In this case the conditions of formation of the adsorption layer were considerably more equilibrium, than in air. The comparison of results in columns I and II

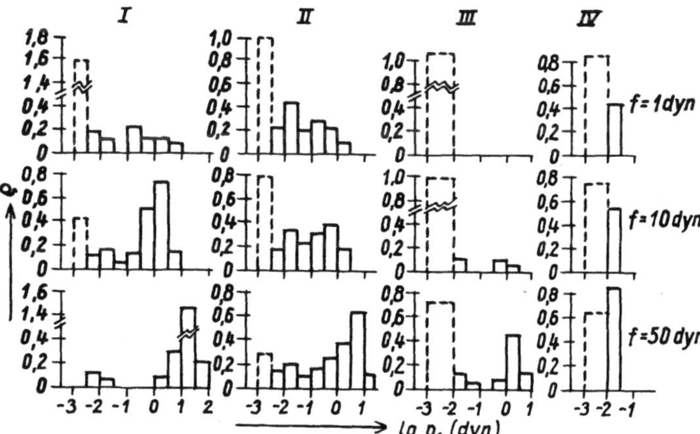

Fig. 3. Histograms of the strength of the contacts, formed by compression of the crystals

   I AgCl in air
  II AgCl + ODA in air
 III AgCl + ODA in heptane
 IV AgCl + ODA in water
ODA: octadecylamine

shows, that when the absorption layer is present on AgCl surface, the second maximum on the histograms, which corresponds to the formation of phase contacts, appears at much greater forces. As for the conditions, when only phase contacts appear between pure AgCl crystals, the compression being equal to or more than 50 dyne (lower figure in column I), a considerable part of weak contacts is still preserved in presence of octadecylamine (lower figure in column III).

Formation of cohesion (phase) contacts is impossible without destruction of the absorption layer. That is why the hindering of formation of strong phase contacts is a direct manifestation of mechanical strength of surfactants adsorption layers. It is known (9, 10) that such layers in the boundary state on the surface of solids become able to endure great mechanical stresses. And though the precise mechanism of effect of absorption layers is not yet clear, the importance of their mechanical strength is undoubtful.

The given method allows to establish the dependence of mechanical properties of surfactant adsorption layers on the thickness of the layer, on chemical nature of surface-active molecules, and on the character of the environment. Fig. 4 presents the comparison of values of

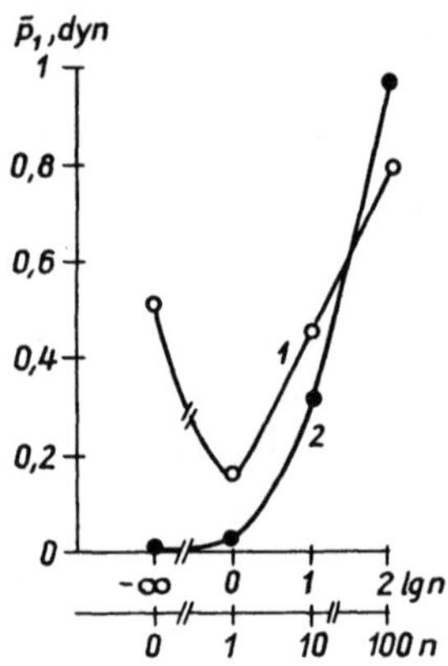

Fig. 4. Dependence of values of average contacts strength (p_1) between. AgCl crystals (1) and samples of glass (2) on the number of octadecylamine layer (n)

average contact strength between AgCl crystals (curve 1) and between samples of glass (curve 2) in presence of octadecylamine layers of various thickness. Under these conditions no phase contacts appear between the particles of glass, since the material is brittle-elastic and gives no plastic deformations. As it is seen from this picture, the ODA monolayer hinders the forma-

tion of phase contacts between samples of plastic AgCl: the average strength of the contacts decreases. The same ODA monolayer does not change the strength of contacts between elastic glass samples. The thickness of the layer being increased to 10 and 100 monolayers, the contacts strength grows in both cases. This growth of the contacts strength is the result of cohesion of the polymolecular layers themselves when they suffer plastic deformation. This increase in strength is the greater, the thicker is the layer.

Table 2. Dependence of per cent of the phase contacts, formed between AgCl crystals, on the force of compression of the crystals f

f, dyne	0.5	5	50
heptane	14	55	96
DA in heptane	1	30	94
CA in heptane	1	20	65
ODA in heptane	0	3	25

Table 2 presents the results of measurements of contact strength between AgCl crystals in solutions of various surfactants (decyl and cetyl alcohols and octadecylamine in heptane). The results are presented as dependences of per cent of the phase contacts (the ones 0.1 dyne and more strong) on the force of compression of the crystals. The table shows that the protective effect of the adsorption layers grows within the homologous series from the lower homologues to the higher ones. The ODA monolayers protective effect is greater, than the one of alcohols. The comparison of the results of measurements of cohesion forces between AgCl particles in ODA solutions in water and heptane (fig. 3, columns III and IV) indicates that in water the ODA absorption layers are stronger, than in heptane: in water they hinder AgCl cohesion to a greater extent. This is obviously connected with the easier destruction of the surfactant layer in the related hydrocarbon medium, i.e. heptane. The destruction is facilitated due to the screening of interaction between surface-active molecules hydrocarbon radicals by related heptane molecules analogous to the same process in mixed films.

Thus we see that the method considered here may be applied for investigations of absorption layers of surface-active substances. It allows to characterize the mechanical properties of such layers at the solid–gas and solid–liquid interfacial boundaries.

The properties of surfactants absorption layers at the liquid–gas and liquid–liquid interfacial boundaries were studied all-round in detail by scientists of well-known colloid chemistry schools of the USSR *(Rehbinder, Trapeznikov)*, Holland *(Leklem, Van den Tempel)*, GDR (in *Sonntag*'s works), Bulgaria (in *Sheludko*'s works), etc. Reliable and quantitatively proved static and dynamic methods of investigation exist for such interfacial boundaries. Unlike this, there were no methods strict and proved enough for studies of surfactants adsorption layers at the surface of solids. It is connected with that only liquid surfaces due to their molecular smoothness were considered to be suitable objects for such investigations.

We have been speaking so far about measurements of cohesion forces between samples of irregular shape with natural rough surface. The wide dispersion of cohesion forces values allows to find out the unhomogeneity of the structure of the surface, but hinders the obtaining of exact quantitative characteristics of mechanical properties of the adsorption layers. For this one should use samples with surface of molecular smoothness. Exact knowledge of geometric parameters and of elastic constants of the material allow (according to *Hertz*) to evaluate the square of the contact area and the critical stress destroying the adsorption layer. As such samples *Amelina* and *Yaminskiy* had used spheric particles of quartz 1.5—2 mm in diameter and the Pyrex glass with fused (hydrophilic) surface, and also quartz particles with hydrophobized surface (11, 12). Hydrophobization was carried out by means of chemical modifying of the surface of the balls with dimethyldichlorsilane. The measurements of cohesion forces between such particles were conducted in air, in water, in heptane and in solutions of surface-active substances. The force of compression of particles and time of application of the force varied from 10^{-3} to 10^2 dyne and from 10 to 1000 sec, respectively. The results of the measurements showed that the transition to spheric particles with molecularly smooth surface entirely changes the picture observed before. Wide scatter of cohesion forces values, as in case of AgCl crystals, is eliminated. The glass and quartz particles depending on the circumstances either gave no cohesion (i.e. the strength of contacts was 10^{-3} dyn and less), or manifested considerable cohesion (the contacts strength was about 10—30 dyne), contacts with intermediate values of strength did not appear. It should be emphasized that unlike in experiments with plastic AgCl great cohesion forces between the quartz balls (10 dyne and more) are not connected with formation of phase contacts. For plastic AgCl crystals the edges of which are drawn together the contacts 0.1 dyne and more strong are phase contacts. For elastic-brittle and smooth quartz particles of large radius of curvature the contacts 10 and more dyne strong are due to van-der-Waals attraction forces when the particles touch each other, and the ones 10^{-3} dyne strong appear when a layer of the liquid environment is present, like in Derjaguin method of crossed threads (13) and in the Bradley method (14).

Hydrophilic particles being brought together in air and in heptane, their cohesion takes place at once and the value of cohesion forces ($p_1 \sim 30$ dyne in air and $p_1 \sim 15$ dyne in heptane) does not depend on the value of the stress and the time of keeping the particles in contact.

The same particles in aqueous environment display no cohesion ($p_1 \sim 10^{-3}$ dyne) at all the values of compression force and time of application of the force.

Cohesion of hydrophilic particles in the lauric and capronic acids heptane solutions, where the particles had been previously kept for the formation of the equilibrium absorption layer, depends considerably on the value of the force and the time of preliminary compression.

When the pressing forces ($\sim 0{,}5$ dyne) and the times of their application (~ 10 sec) are small, the cohesion of particles does not take place. The increase of forces and the time of its application up to some certain values results in a sharp increase of adhesion forces from 10^{-3} dyne to ~ 10 dyne, i.e. to the values, corresponding to adhesion force between particles in pure heptane. Such stabilizing effect of surfactants, namely the ability of their adsorbed layers to decrease the interaction of hydrophilic particles in nonpolar medium, is connected with formation of mechanical barrier, caused by mechanical properties of adsorbed layers, that is their ability to resist deformation and destruction. It appears that the resistance of capronic and lauric acid layers has viscous character, i.e. they behave as quasi-liquid ones.

In the same experiments with stearic acid there was found no cohesion of the quartz particles up to 50 dyne forces and 1000 sec times of contact. Evaluation according to *Hertz* showed that such layers resist normal stresses of the order of 5×10^8 dyne/cm², i.e. the stresses of the order of strength of molecular crystals.

It is interesting to compare this value with the one of two-dimensional pressure P_s, measured with the Langmuir method. The maximum value of the two-dimensional pressure 30—40 dyne/cm, divided by the thickness of the monolayer, also gives very high values of the critical side compression — about 2×10^8 dyne/cm². Thus the results considered are in accordance with the ones obtained with the Langmuir method and correspond to *Zisman*'s concepts on the character of the aliphatic acids adsorption layers on the surface of solids (15).

Wide-scale variation of the conditions of the experiment, which is possible due to the given method, allows to model contact interactions in various disperse systems and obtain quantitative data necessary for evaluation of nature and degree of stability of disperse systems.

Here are some preliminary data obtained by *Yaminskiy* on investigations of quartz balls hydrophobized with dimethyldichlorsilane cohesion in water (12). Cohesion of such particles in water takes place immediately with a force of 20 dyne. In a non-polar medium-heptane-cohesion (taking into account the possible electrization) was not found at any forces and times of compression. We observe here complete analogy with the above results of experiments on bringing together hydrophilic particles in water. These observations well agree with the stability of suspensions of hydrophilic powders in water and hydrophobic ones in oil and vice versa: strongly pronounced coagulation of the first in oil and of the second in water.

The addition of the four first normal aliphatic alcohols (from methyl to butyl), to water and also of micelle-forming surfactants (sodium dodecyl sulfate, cetyl pyridinium bromide and polyoxyethylene dodecyl ester) resulted in decrease of cohesion forces between hydrophobized beads. This decrease was the greater, the higher was the concentration of surfactant in the solution and its position in the homologous series (for the alcohols). *In this case the value of cohesion forces did not depend on the force and time of compressing.*

On the grounds of these measurements the values of specific particle interaction free energies have been calculated in accordance with *Deryaguin*'s theory (16) of adhesion of elastic particles with convex surfaces. These results were compared with the ones of surface tension measurements of the surfactants under investigation at the interface of paraffine wax and ethyl alcohol solutions. In the latter case surface tension values were obtained from *Dann*'s publication (17). In this work surface tension was determined from contact angle measurements. The comparison of concentration dependence of adhesion energies with surface tension isotherms reveals that they have many common features in absolute energy values and in their relative dependences.

In case of sodium dodecyl sulphate concentration dependence of adhesion energy was compared with values of surface tension decrease at surfactant solution — methylated glass interface (in the concentration region before critical micelle concentration, CMC). The latter dependence was calculated from the adsorption measurement data of sodium dodecyl sulphate at the methylated glass surface.

Fig. 5 represents the results of such comparison. It is seen from the picture that both curves coincide. It shows, that in contrast with nonpolar medium, where the stabilizing effect of adsorbed surfactant layers has the nature of a mechanical barrier, in aqueous medium the decrease of hydrophobic particles interaction energy is a result of surface tension decrease at solid-solution interface.

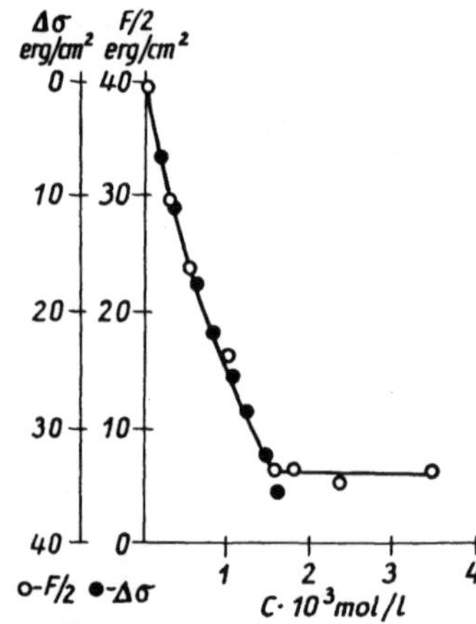

Fig. 5. Concentration dependence of adhesion energies (F/2) of hydrophobizated particles in water solution of sodium dodecyl sulphate and isoform of surface tension ($\Delta\sigma$) lowering particle — water solution

Thus the method allows to reveal different mechanisms of the stabilizing effect of surface-active substances as depending on particles surface nature and dispersion medium nature for disperse systems with solid disperse phase.

The described mehtod of measurements of cohesion forces in contacts between separate crystals was used by *Amelina* and *Vaganov* for an experimental investigation of regularities of formation of crystallization contacts as crystals agglutinate in the process of the new phase formation (18, 19). Appearance of such contacts is the basis of crystallization disperse structures development in the process of hardening of mineral astringents. The regularities of formation of such structures were studied in detail in works of *Rehbinder, Segalova* et al. (20); *Pollack* (21) had offered the theoretical description of separate stages of the crystallization structure formation process. These works formed the grounds of the physico-chemical theory of hardening of mineral astringents. According to this theory the emerging of crystallization structures takes place only under certain conditions, that is when supersaturations high enough are present in the liquid phase of the hardening suspension and when the mobility of the appearing crystals is limited.

Experimental works in this field were based upon macroscopic measurements, i.e. determining of strength of materials, appearing as a result of agglutination of separate crystals of the new disperse phase in the process of condensation (crystallization) from solutions. Along with that the problem of physico-chemical regularities of emergence of the main strength carriers in disperse structures, namely the crystallization contacts, that is the problem of the micromechanism of structures formation remained unsolved, as direct experimental investigations of separate crystals growing together in the process of appearance of the new phase were absent.

The essence of the experiments in our recent investigations was as follows: two gypsum crystals ($5 \times 5 \times 0.5$ mm in size) cleaved out of a monocrystal were placed into calcium sulphate solutions of various concentrations and were brought in touch with the given force "*f*". The crystals had stayed for a certain time in the fixed position and were separated afterwards. The force was measured, required for the

separation, i.e. the strength of an individual contact (p_1). Each series of the experiments included no less than 100 measurements and was carried out with the same couple of crystals. The geometry of the contact corresponded to the mutual position of the crystals ("edge to edge"); the crystals touched each other with the edges of certain symbols. Supersaturated solutions were prepared by means of calcium sulphate semihydrate, which is more soluble, than gypsum, dissolving in water, thorough filtration and diluting of the original solution to the required concentration. It was found, that during an experiment the concentration of the solution did not change. The time of contacting (t) varied from 10 to 1000 sec, the supersaturation of the solution $\alpha = C/C_0$ — from 1 to 3, the forces at which the crystals were brought in touch "*f*" — from 0.1 to 10 dyne (C and C_0 are concentrations of supersaturated and saturated with respect to gypsum solutions respectively).

Experimental results, as in the experiments considered above, were presented as differential histograms of contacts strength p_1 logarithms. In these experiments an extremely sharp leap-like transition is observed from coagulation — type contacts (with strength 10^{-3} dyne and less) to contacts of crystallization type (with strength 10^{-1} dyne and more), contacts with the strength in the interval between 0.01 and 0.1 dyne not being found in any of the experimental series. The share of crystallization contacts W_k, i.e. the ratio of the number of emerging strong contacts w_t with $p_1 \geq 0.1$ dyne to the total number of the registered contacts w_0 ($W_k = w_t/w_0$) grows with the increase in supersaturation and time of contacting.

The dependence observed allows to compare the emerging of crystallization contacts with the initial stage of the crystals' agglutination process, i.e. the fluctuational formation of a nucleus-contact critical for the given super-saturation, which serves as the first crystallization "bridge". The presence of a marked leap-like transition from one type of contacts to the other gives an opportunity to register the fact of intergrowth itself and to establish due to that the quantitative dependence between the share of crystallization contacts W_k, i.e. the probability of growing together, and the parameters varied, this is clearing up certain features of an elementary act of agglutination.

The analysis of the experimental data had shown, that the dependence of probability of agglutination on time may be presented as

$$W_k = 1 - \exp(-It) \tag{1}$$

where I is a value reverse to the characteristical time during which the crystals agglutinate in 63% of cases from the total number of experiments W_0 (i.e. I_0 is the rate of agglutination).

According to well-known concepts of the fluctuation theory

$$I = I_0 \exp(-A_k/kT), \tag{2}$$

where the work of formation of the nucleus

$$A_k = \frac{4h^2\sigma^2}{h\dfrac{kT}{v}\ln\alpha + 2\sigma}$$

where h is the height of the nucleus, v is the molecular volume, σ the specific surface energy at the boundary of the nucleus with the solution (21).

Values of I, obtained on the grounds of experimental data from the eq. [1] at various values of t were used for evaluation of the work of the nucleus-contact formation A_k (table 3). The obtained values of A_k are quite reasonable from the point of view of the fluctuation theory of new phase formation; their comparison with the values of work of formation of two-dimensional gypsum nuclei (21) points out that namely contact nucleus is thermodynamically expedient. Supposing that the nucleus contact has the shape of a prism with the height h and the basement a_k we are getting the values of $a_k = 9 - 6$ Å, at $\alpha = 1 - 3$ and $h = 10$ Å.

It is known that emerging of crystallization contacts requires a combination of certain supersaturation in the environment and mechanical forces keeping the crystals in a definite

Table 3. Values of the work of formation of a nucleus-contact (A_k) at various degrees of supersaturation α and forces of mechanical compression f

f, dyne	α	A_k/kT
0.1	1.2	6.7
	1.5	5.9
	1.8	5.3
	3.0	4.3
1	1.2	8.0
	1.5	6.8
	1.8	6.1
	3.0	48
10	1.2	6.7
	1.5	5.9
	1.8	5.3
	3.0	4.3

position so that they are fixed in respect with each other. Such forces may be either a result of external pressure or a result of pressure developing in the process of directed growth of the crystals inside the crystallization network already formed. In order to clear up the role of mechanical forces in the process of formation of crystallization contacts we have conducted investigations of dependence of probability of growing together on the value of the force of compression of the agglutination crystals.

The results of such investigations (table 4) showed that the probability of agglutination of crystals increases as the force bringing them together increases at all values of α and t. The analysis of the data obtained allowed to connect this increase with the increase in the factor I_0 in the expression (2), which should include the size of the active area where the emerging of the contact nucleus is possible. The corresponding square may be evaluated through the yield point τ^* ($s \sim f/\tau^*$) in case of plastic deformation of crystals being compressed or according to *Hertz* if the deformation is elastic.

Table 4. Dependence of probability of intergrowth of gypsum crystals (%) on the value of mechanical forces f, at which they are brought together in solutions with different degree of supersaturation at different times of contact t

$\alpha = C/C_0$	1.2			1.5			1.8			3.0		
f, dyne / t, sec.	0.1	1	10	0.1	1	10	0.1	1	10	0.1	1	10
10	0	0	0	0	1	5	0	2	7	0	7	38
100	0	4	17	0	8	24	1	14	49	4	57	83
1000	4	22	33	7	44	68	10	60	96	33	91	100

Thus the analysis of the experimental data (taking into account the possible influence of mechanical forces on the degree of supersaturation (19)) allowed to determine the functional dependence of a probability of agglutination of crystals (emerging of crystallization contacts between them) in the process of appearance of a new phase on such main physico-chemical parameters, as the value of supersaturation, time of contacting, force of mechanical compression, which in the first place determine the entire process of crystallization hardening structures formation, as

$$W_k = 1 - \exp\left\{ - \varkappa f/\tau^* \cdot \exp\left[\frac{4h^2\sigma^2}{kT\left\{h\dfrac{kT}{v}\left(\ln\alpha - \dfrac{\Pi v}{kT}\right) + \sigma^*\right\}}\right]\right\} \cdot t$$

where $\varkappa$ is the factor having dimensions of $\sec^{-1}\,\mathrm{sm}^{-2}$,

v is the molecular volume,

σ is the specific surface energy of the nucleus-contact at the boundary between the nucleus and the solution,

Π is the stress at the considered area under the action of the applied compression forces,

$\dfrac{\Pi}{kT}$ is the correction factor connected with changing of supersaturation under the influence of the stress applied,

$\sigma^* = 2\sigma - \sigma_g$ is the algebraic sum of specific surface energies of elements of the interfacial boundary appearing and disappearing simultaneously as a result of the nucleus-contact emerging.

σ_g is the specific surface energy at nucleus-crystal interface.

The meaning of f, τ^*, α, t was indicated above.

In this report we have tried only to use the analysis of contact interactions cohesion forces between particles in disperse systems with the solid disperse phase — as an actual example of application of basic directions of modern colloid chemistry. These directions are: the analysis of mechanism of effect of surfactants, studies of stability and structure formation in disperse systems.

In fact, all these directions are reflected in investigations of various aspects of interactions between particles in disperse systems. In its turn, the comprehensive investigation of contact interactions appears to be very fruitful from the point of view of working out the main directions of modern colloid chemistry.

References

1) *Shchukin, E. D., R. K. Yusupov, E. A. Amelina,* and *P. A. Rehbinder,* Kolloid, zh. **41**, 913 (1968).
2) *Shchukin, E. D., E. A. Amelina, R. K. Yusupov,* and *P. A. Rehbinder,* Dokl. AN SSSR (Dokl. Akad. Nauk SSSR) **170**, 1037 (1970).
3) *Yusupov, R. K., E. A. Amelina, E. D. Shchukin,* and *P. A. Rehbinder,* Dokl. Akad. Nauk SSSR **200**, 1077 (1971).
4) *Yusupov, R. K., E. A. Amelina,* and *E. D. Shchukin,* Kolloid, zh. (in print).
5) *Babak, V. G., E. D. Shchukin, E. A. Amelina* and *P. A. Rehbinder,* Dokl. Akad. Nauk SSSR **206**, 132 (1972).
6) *Amelina, E. A., R. K. Yusupov,* and *E. D. Shchukin,* Collection: Investigations on physic-chemistry of contact interactions, p. 110 (1972).
7) *Rehbinder, P. A., E. D. Shchukin,* and *L. Ya. Margolis,* Dokl. Akad. Nauk SSSR **154**, 695 (1964).
8) *Amelina, E. A., R. K. Yusupov,* and *E. D. Shchukin,* Kolloid, zh. **36**, N 5, (1974).
9) *Hardy, W. B.,* Collected Scientific Papers (Cambridge 1936).
10) *Zisman, W.,* Friction and Wear (Amsterdam 1959).
11) *Yaminskiy, V. V., R. K. Yusupov, E. A. Amelina, V. A. Pchelin,* and *E. D. Shchukin,* Kolloid, zh. (in print).
12) *Yaminsky, V. V., E. A. Amelina,* and *E. D. Shchukin,* Kolloid, zh. (in print).
13) *Malkina, A. D.* and *B. V. Derjaguin,* Kolloid, zh. **12**, 431 (1950).
14) *Bradley, R. S.,* Phil. Mag., Ser. **7**, **13**, 953 (1932).
15) *Levine, O.* and *W. A. Zisman,* J. Phys. Chem. **61**, 1188 (1957).
16) *Derjaguin, B. V., N. A. Krotova,* and *V. P. Smilga,* Adhesion of Solids, Nauka (Moscow 1973).
17) *Dann, Y. R.,* J. Colloid Interface Sci. **32**, 302 (1970).
18) *Shchukin, E. D., E. A. Amelina, R. K. Yusupov, V. P. Vaganov,* and *P. A. Rehbinder,* Dokl. Akad. Nauk SSSR **213**, 155 (1973).
19) *Shchukin, E. D., E. A. Amelina, R. K. Yusupov, V. P. Vaganov,* and *P. A. Rehbinder,* Dokl. Akad. Nauk SSSR **213**, 398 (1973).
20) *Segalova, E. E.* and *P. A. Rehbinder,* Collection: The New in Chemistry and Technology of Cement, p. 232 (Moscow 1962).
21) *Pollack A. F.,* Hardening of Monomineral Astringents (Moscow 1966).

Authors' address:

E. D. Shchukin and *E. A. Amelina*
Chair of Colloid Chemistry
Moscow State University Moscow, 117234 (USSR)
Institute of Physical Chemistry
Academy of Sciences of the USSR Moscow

Progr. Colloid & Polymer Sci. **61**, 80–86 (1976)
© 1976 by Dr. Dietrich Steinkopff Verlag GmbH & Co. KG, Darmstadt
ISSN 0340-255 X
Plenary lecture of the IUPAC-Conference on Colloid and Surface Science in
Budapest, September 15–20, 1975

Department of Chemistry, Faculty of Engineering, Yokohama National University, Ooka, Minamiku,
Yokohama (Japan)

Dissolution due to the orientation, arrangement and structure formation of molecules

K. Shinoda

With 9 figures and 1 table

(Received December 9, 1975)

Introduction

It is my high honor to participate in the International Conference on Colloid and Surface Science (Hungary 1975) by presenting the Main Lecture on Colloidal Solutions.

There are many types of solutions, among which regular solutions, electrolyte solutions and (regular) polymer solutions have been well investigated by respective ingenious scientists (1, 2, 3). The types of solvents and the intermolecular forces involved in these solutions differ from hydrocarbon to water and from London dispersion force to Coulomb force. It appears that these solutions cover almost all kinds of solutions. Nevertheless, solvent and solute molecules randomly mix by thermal motion in these solutions and the theory also assumes completely random mixing.

These idealized solutions are not general, but rather exceptional among so many solutions which we encounter in the study of biological systems, in the process of manufacturing or in environmental problems. Most of my works have been directed to apply and extend solution thermodynamics to practically important yet unexplored systems, such as aqueous nonelectrolyte solutions, surfactant solutions, solubilized solutions, water-soluble polymer solutions and so on. The most striking feature of these solutions may be the dissolution due to the orientation, arrangement, conformation and structure formation of molecules, which are otherwise practically insoluble by random mixing only. I would like to designate these solutions, which are mostly colloidal, as advanced solutions.

Table 1. See text

Solutions in which molecules are randomly mixed	Solutions in which orientation, arrangement and structure formation of molecules is indispensable for dissolution
Athermal solutions	Chelate solutions
Regular solutions	Surfactant solutions
Electrolyte solutions	Liquid crystals, gels
(Regular) polymer solutions	Solubilized solutions
	Bio-colloids, water soluble polymer solutions

The mechanisms of dissolution of surfactants and water soluble polymers in water will be described as typical examples.

Solubility of surfactant in water

Surface active substances dissolve in water in singly dispersed state at very low concentration. If the concentration exceeds the saturation concentration, i.e., critical micelle concentration (c.m.c.), aggregated particles called micelles are formed.

The surface tension of aqueous surfactant solution decreases up to the saturation concentration of singly dispersed species and then stays nearly constant with the concentration as shown in fig. 1.

On the other hand, surface excess Γ_2 of surfactant above the c.m.c. revealed by radio tracer measurements (4) showed a large constant value. Applying Gibbs' adsorption isotherm,

$$\left(\frac{\partial \gamma}{\partial \ln c_2} \right)_T \left(\frac{\partial \ln c_2}{\partial \ln a_2} \right) = - RT\, \Gamma_2 \qquad [1]$$

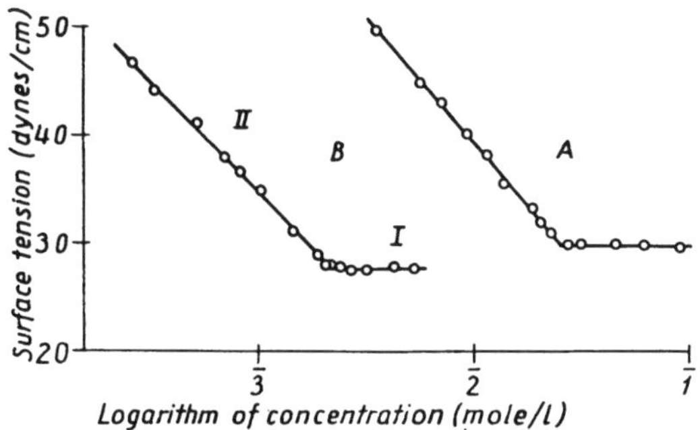

Fig. 1. The surface tension in aqueous solutions of octyl (A) and decyl (B) glucosides as a function of the logarithm of concentration at 25 °C

to these facts, it is concluded that the activity of surfactant does stay nearly constant above the c.m.c., and therefore the micelle formation is akin to a phase separation (5). The aggregation number of surfactant per micelle is not infinite but finite thus we designate it as pseudo-phase. *Hall* and *Pethica* applied the thermodynamics of small systems to micellar solutions (6). Thermodynamic functions, such as partial molar free energy, enthalpy, volume etc. stay nearly constant with the concentration, but the solution is transparent one phase. This means practically infinite solubility. This is a revolutional fact to attract solution scientists. A novel, unique mechanism of dissolution had to be proposed in order to understand this phenomenon. Let us consider the solubility of dodecanol in water. Dodecanol will dissolve in water in molecularly dispersed state up to the solubility, and then excess solid or liquid dodecanol will separate above solubility. Broad curve illustrates the solubility curve of dodecanol (ordinary substances). Below the melting point of dodecanol

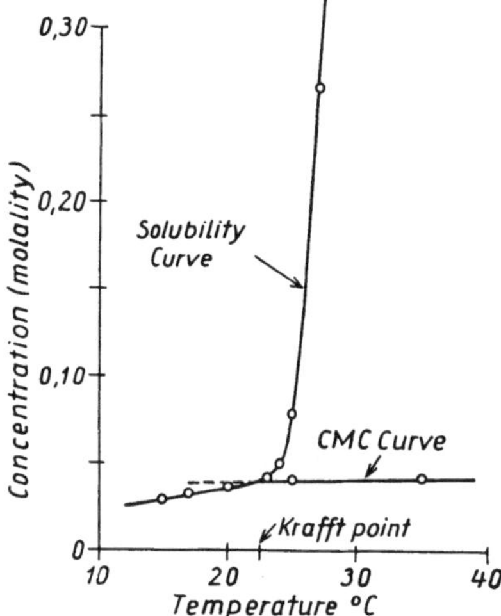

Fig. 3. Phase diagram of sodium decyl sulfonate-water system close to the Krafft point

in the presence of water, solid dodecanol separates from solution and above it liquid dodecanol separates. In the case of ionic surfactant-water system, the situation changes due to the micelle formation. Surfactant dissolves in singly dispersed state up to the saturation concentration above which hydrated solid surfactant phase separates below the melting point (Krafft point) of hydrated surfactant and micelles will be formed above the Krafft point. As the size of micelles is so small compared with the wave length of light the solution is transparent one phase. Solute-solid and liquid-(micelle)-solid equilibrium curves constitute solubility curve. The physical meaning of the solubility curve is different of that of ordinary compounds. This is the reason why the solubility of ionic surfactant abruptly increases above the Krafft point (5).

Due to our common sense, paraffin chain compounds may scarcely dissolve in water. This reasoning is manifested in the very small c.m.c. We often experience a small fragment of used up soap in the sink does not disappear over a week. In order to increase the solubility the Krafft point should be depressed or the temperature has to be raised above the Krafft point. Any attempt to enhance the saturation concentration of singly dispersed molecules, i.e., dissolution by random mixing, is meaningless to increase solubility in association colloidal solutions. Micellar dispersion due to the orientation, arrangement and structure formation of molecules is responsible to such properties.

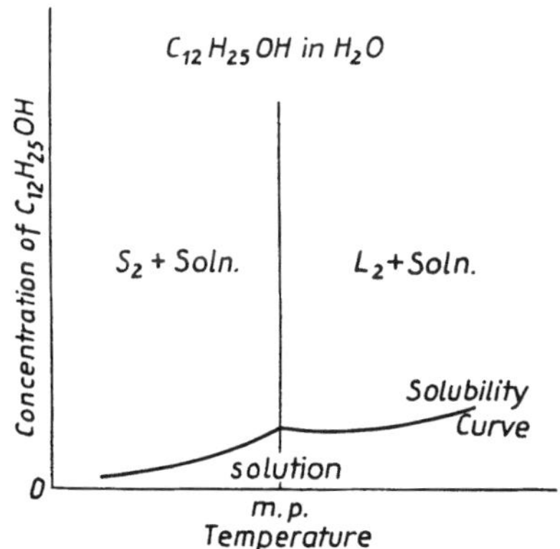

Fig. 2. Schematic diagram of the solubility of dodecanol in water as a function of temperature

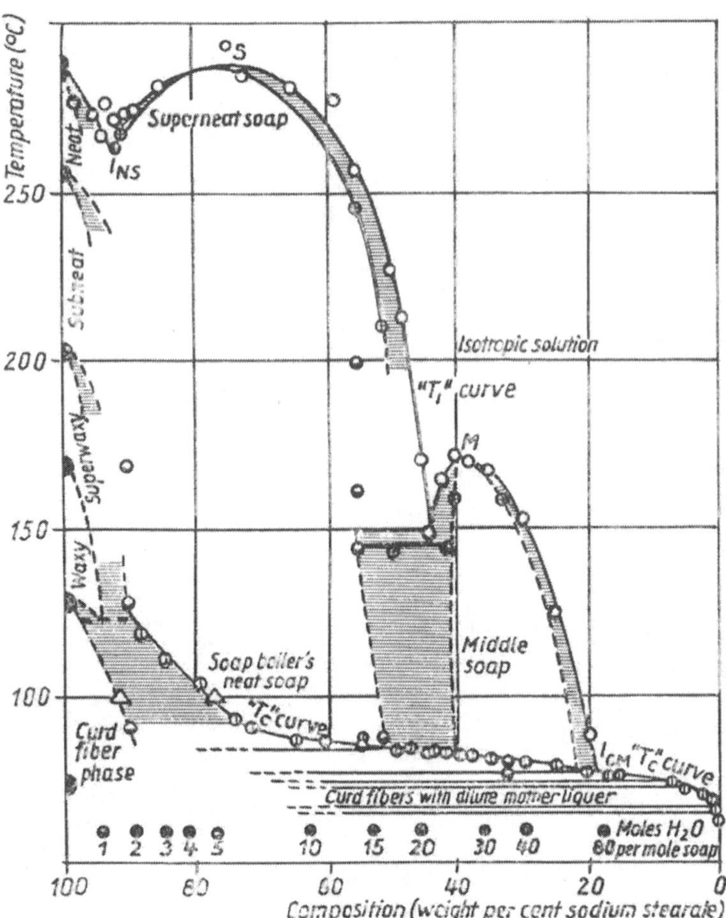

Fig. 4. Phase diagram of soap-water system (7)

The fraction of micelles against bulk solution increases with the concentration of surfactant and finally the whole system becomes liquid crystalline phase or (hydrated) solid phase as shown in fig. 4 (7).

In other expression, water will dissolve into surfactant and form liquid crystalline phase at first. Water swollen liquid crystalline phase will disperse as bimolecular leaflets (micelles) into water with the further addition of water.

In the case of nonionic surfactants, the melting point in the presence of water is usually below 0 °C. And, we might expect that nonionic surfactant will mix with water over almost any composition just as ionic surfactants above their Krafft points. As the hydrophilic property of ionic polar group is strong, 1:1 type ionic surfactants usually form micelles above the c.m.c. If the hydrophilic property of nonionic surfactant is strong enough, or in other words, if water dissolves infinitely in nonionic surfactant phase, it forms micelle particles in water just as

ionics do. If the hydrophilic property is not strong, it aggregates infinitely above the solubility and splits into two phases just as octanol or dodecanol does.

Nonionic surfactant, such as octaoxyethylene dodecylether, dissolves in water forming micelles at low temperature. If the temperature of the solution is raised to the cloud point, it suddenly becomes cloudy and the solution splits into two phases as shown in fig. 5 (8). One is a water phase containing molecularly dispersed nonionic surfactant (~ 0.02 wt%), and the other is a surfactant phase water dissolved in it. Thus, it is evident that the cloud point curve is a liquid-liquid solubility curve. Part A—B of the solubility curve is extremely close to the water axis, which indicates the very small solubility of paraffin chain compounds in water. The physical meaning of curve A—B is the same as that of an ordinary solubility curve.

On the contrary, the solubility of water in nonionic surfactant phase is fairly large, i.e.,

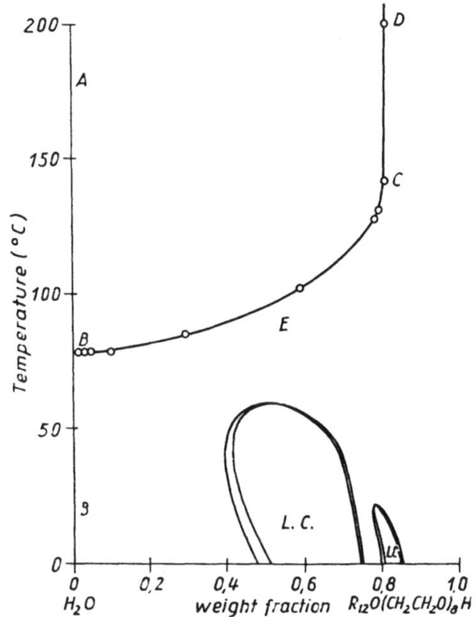

Fig. 5. Phase diagram of $C_{12}H_{25}O(CH_2CH_2O)_8H-H_2O$ system

about 19 wt% (87 mole%) because the molar volume of water is small and enthalpy of solution per molecule is small. At the lower temperature the hydration between water and the oxygen atoms of the hydrophilic group increases, and the solubility of water in surfactant may further increase. Surfactant will have a layered structure as in the neat phase. Water and surfactant may be both continuous in this composition range. If more water is dissolved, the surfactant phase may disperse in the form of bimolecular leaflets, i.e., micelles. If such pseudo-phase inversion on the bimolecular leaflet scale occurs, the surfactant phase will mix with water phase infinitely. Namely, the solution between B and E below the cloud point is considered as a mixture of the water phase and hydrated surfactant phase. Above this curve (temperature) surfactant aggregates infinitely, but below this curve the surfactant aggregates finitely and disperses as stable colloids. In other words, the solubility of water in surfactant phase is finite until surfactant aggregate infinitely, but it is infinite below cloud point. The physical meanings of curves AB and BE are different and two curves intersect discontinuously and perpendicularly. Structure and property of solution close to C may not change appreciably with temperature decrease. Actually, the relative activity of water does not change over wide

volume fraction range (9). Applying the Gibbs-Duhem relation,

$$X_1\,d\mu_1 + X_2\,d\mu_2 = 0 \quad (P.T. = const) \qquad [2]$$

we can conclude that the relative activity of surfactant does not change also over the same volume fraction range B to C. This means the micelle formation is a phenomenon similar to a phase separation and the thermodynamic properties of micellar solution resemble to two phases mixture.

Solution of water-soluble polymers (partially saponified polyvinyl acetate).

It is considered that partially saponified polyvinyl acetate (PVA-Ac) resembles the block copolymers of vinylacetate and vinylalcohol. The phase diagram of $PVA\text{-}Ac-H_2O$ as a function of temperature is shown in fig. 6 (10). The curve ABCD represents the mutual solubility of water and PVA-Ac. It is clear from fig. 6 that the liquid-liquid solubility curve is affected sensitively to the saponification degree, but not to the polymerization degree. Although the solution is so viscous and difficult to dissolve above 15 to 25 wt% (freeze-dry PVA-Ac is easier to dissolve), it is soluble in water over all composition below the cloud point, B—C. The solution becomes

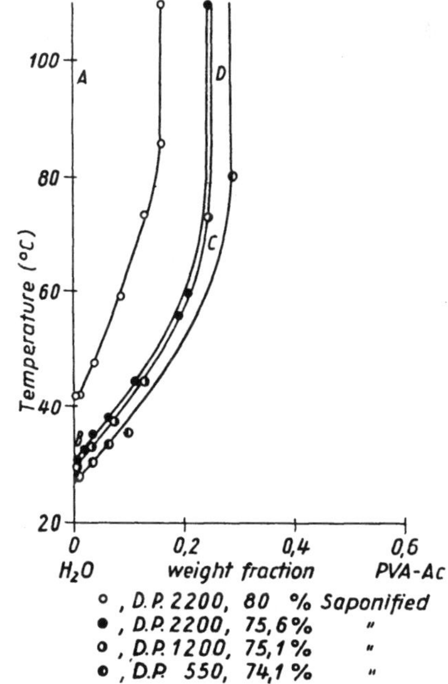

Fig. 6. Phase diagrams of H_2O-PVA-Ac of different saponification degree (D.S.) and polymerization degree (D.P.).

cloudy and splits into two phases at temperature above the cloud point curve. The one is a polymer phase containing a large amount of water, and the other is the water phase containing practically no polymer. The activity of water in this phase is close to unity as shown later. The steep slope of BC curve may result from the distribution of the saponification degree of polymers.

The solubility of polymer in water phase is conventionally shown by

$$\ln a_2 = \ln \Phi_2 + \Phi_1(1 - v_2/v_1) \\ + v_2 \Phi_1^2 B'/RT \, . \qquad [3]$$

On the other hand, the solubility of water in polymer phase is conventionally shown by

$$\ln a_1 = \ln \Phi_1 + \Phi_2(1 - v_1/v_2) \\ + v_1 \Phi_2^2 B'/RT \qquad [4]$$

where a is the relative activity, v_i the molecular volumes of respective components in solution, Φ_i the volume fractions and B' the enthalpy of solution per unit volume (3). It is expected from eq. [3] that the solubility of polymer may be very small, because the molecular volume v_2 and therefore the heat of solution of polymer is very large. Inversely, the molar volume of solvent v_1 is so small that water dissolves infinitely in polymer, provided the heat of solution per mole is smaller than $\frac{1}{2}RT$.

The cloud point curve BC intersects with the solubility curve AB. Similar discontinuity in solubility curve is observed in nonionic surfactant-water system (8, 9). In the case of polymer, curve AB approaches to the water axis much closer than nonionic surfactant solution, because polymer does not dissociate into small molecules. Whereas water phase contains the saturation concentration of surfactant molecules in the case of association colloids. The solubility of water in polymer phase is large (99,99 mole% or $\sim$ 76 wt%) already at point C. The dissolution of water into polymer proceeds with temperature depression due to the increase of ice-berg formation and hydration of water surrounding solute molecule, and finally water and polymer mix each other completely below the temperature at B. The schematic diagram of the dissolution process of polymer which consists of hydrophilic and lyophilic groups, in water is shown in fig. 7 (10).

It is considered that water dissolves infinitely into polymer phase and pseudo-phase inversion

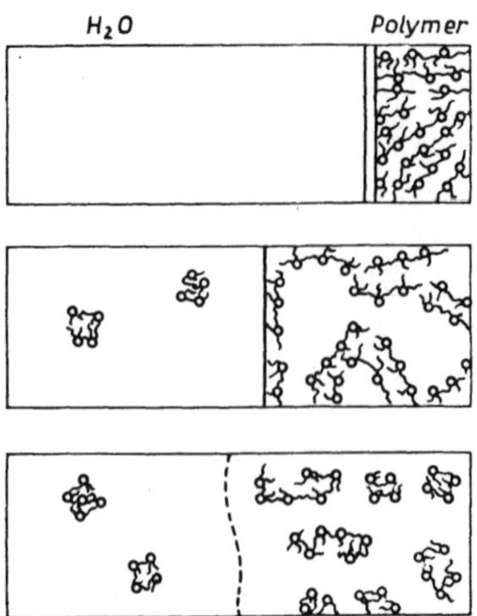

Fig. 7. Schematic diagram of dissolution process of polymer in water which consists of hydrophilic and lyophilic groups. 1 Before mixing, 2 swelling in polymer phase, 3 pseudophase inversion in polymer phase

occurs in polymer phase. If it happens water becomes continuous phase. Polymer molecules will orient, rearrange so as to decrease the free energy of mixing. This phenomenon is akin to the micellar dispersion of surfactant due to the increase of water solubility into surfactant phase at low temperature. If the saponification degree of PVA-Ac is smaller, i.e., less hydrophilic, BC curve will shift to lower temperature and CD curve will shift to higher concentration, because the hydration of water per unit weight of polymer will decrease. Only swelling of water in polymer may occur below a certain saponification degree ($\sim 70\%$), and insoluble in water. Although molecules are aggregated to micelles by physical bond in the case of surfactant, and monomers are linked by covalent bond in polymer, the mechanism of mixing with water is similar. Important conditions for complete dissolution of these substances are 1) liquid state of polymer in the presence of solvent, 2) the finite aggregation of these molecules in water due to the orientation and arrangement, and 3) infinite mixing with water.

The relative activity of water in aqueous solution of PVA-Ac at 25 °C is plotted in fig. 8 (10).

Because of the large molecular weight, polymer solution whose weight per cent ranges from 0 to 50 wt% corresponds to 0—0.00015 in mole fraction unit. The dotted line expresses the relative activity of water in ideal dilute solution. The curve represents the relative activity of

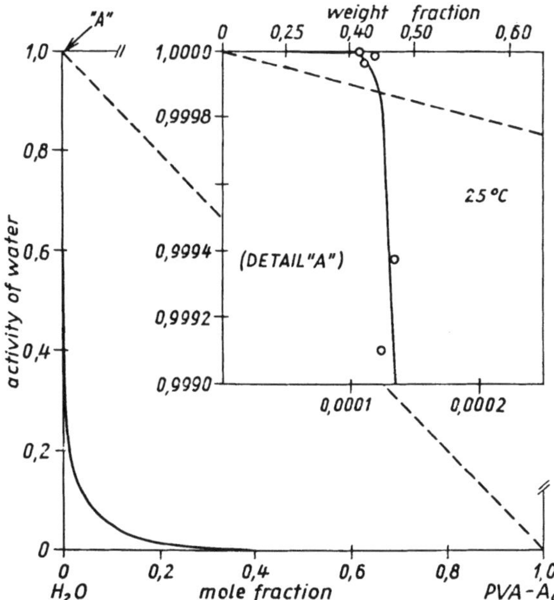

Fig. 8. Relative activity of water in aqueous solution of PVA-Ac (polymerization degree, 2200; 75.6% saponified) at 25 °C

water determined by isopiestic method. The relative activity of water is practically equal to 1 up to 40 wt% solution, i.e., deviates to positive side from that of ideal solution. Applying Gibbs-Duhem equation,

$$X_1\left(\frac{\partial \ln a_1}{\partial X_1}\right) + X_2\left(\frac{\partial \ln a_2}{\partial X_1}\right) = 0. \qquad [5]$$

We know that the activity of polymer also does not change with concentration. The relative activity of water decreases rapidly if the concentration exceeds 40 wt%. It can be concluded from this result that the solution is akin to two phases mixture of pure water and water swollen polymer phase. Hence, we know one phase solution below the cloud point (25 °C) and two phases solution above the cloud point (60 °C) in fig. 6 resemble each other thermodynamically, but not optically. The difference between two states is that the hydrated polymer aggregates infinitely above the cloud point, whereas the aggregation is finite below the cloud point. The mutual solubility of liquids in ordinary solution changes gradually with temperature. On the contrary, polymer (PVA-Ac) is practically insoluble in water above the cloud point, but it mixes with water over all compositions slightly below the cloud point. It is concluded that the change from the complete solubility to insolu-

bility with the small change of temperature is characteristic to polymers. Similar phenomenon can be observed also to the minute change in hydrophile-lypophile balance of polymer molecules, because the cloud point of PVA-Ac is shifted to higher or lower temperature depending on the saponification degree. It is also depressed or raised in the presence of third substances. The effect of added organic additives on the cloud point of 2 wt% aqueous solution of PVA-Ac (saponification degree 75.6% and polymerization degree 2200) is shown in fig. 9.

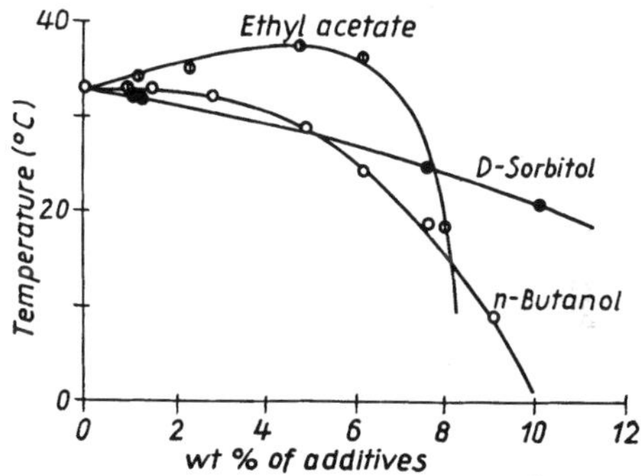

Fig. 9. Effect of organic additives on the cloud point of 2 wt% aqueous solutions of PVA-Ac (saponification degree, 75.6%; polymerization degree, 2200)

It is then concluded that the change from complete solubility to insolubility of polymer in water may occur by the addition of third substances, for example, sorbitol at a constant temperature. The importance of this finding is that the similar phenomenon will occur in biological systems which may explain the consumption or accumulation of fatty substances.

Summary

Pioneering works on regular solutions, electrolyte solutions and (regular) polymer solutions have been done already before 1953. The kinds of intermolecular forces involved and the types of solvents treated in these solutions seem to cover almost all kinds of solutions. Nevertheless, solute and solvent molecules randomly mix by thermal motion in these solutions and the theory also assumes random mixing. These idealized solutions are not general, but rather exceptional among so many solutions which we encounter in our research, i.e., surfactant solutions, biological systems, solutions in the process of manufacturing or in environmental problems, etc. Hence, the study of solutions should be directed now to the more sophisticated, delicate and realistic systems which are not yet well developed. The most striking feature of these solutions may be the dissolution due to the orientation, arrangement, conformation and structure formation of molecules, which are otherwise practically insoluble by random mixing only. The mechanisms of dissolution of micellar solutions of surfactant and water soluble polymers were explained as typical examples. The solubility of these substances in water is very small, but that of water in these substances is infinitely large. The solution resembles to two phases mixture of water and water swollen solute phase for both cases over wide concentration range.

References

1) *Hildebrand, J. H.* and *R. L. Scott*, Regular Solutions (New Jersey 1962); *Hildebrand, J. H., Prausnitz*, and *R. L. Scott*, Regular and Related Solutions (New York 1970).
2) *Debye, P.* and *E. Hückel*, Physik. Z. **24**, 185 (1923); *Harned, H. S.* and *B. B. Owen*, Physical Chemistry of Electrolyte Solutions, 3rd Ed. (1957).
3) *Flory, P. J.*, Principles of Polymer Chemistry (Cornell 1953).
4) *Judson, C. M., A. A. Argyle, J. K. Dixon,* and *D. J. Salley*, J. Chem. Physics 18, 1302 (1950); *Nilsson, G.*, J. Phys. Chem. **61**, 1135 (1957); *Tajima, K., M. Muramatsu,* and *T. Sasaki*, Bull. Chem. Soc. Japan **43**, 1991 (1970).
5) *Shinoda, K.*, Chapter 1 in: Colloidal Surfactants, pp. 6—8 (New York 1963); *Shinoda, K.* and *E. Hutchinson*, J. Phys. Chem. **66**, 577 (1962).
6) *Hall, D. G.* and *B. A. Pethica*, in: Nonionic Surfactants, pp. 516—557 (New York 1967).
7) *McBain, J. W., R. D. Vold,* and *M. Frick*, J. Phys. Chem. **44**, 1013 (1940).
8) *Shinoda, K.*, J. Colloid Interface Sci. **34**, 278 (1970).
9) *Clunie, J. S., J. F. Goodman,* and *P. C. Symons*, Trans. Faraday Soc. **65**, 287 (1969).
10) *Kunieda, H.* and *K. Shinoda*, ACS Symposium Series No. 9, 278 (1975).

Author's address:

Kozo Shinoda
Department of Chemistry
Faculty of Engineering
Yokohama National University
Ooka-2, Minamiku, Yokohama (Japan)

Progr. Colloid & Polymer Sci. **61**, 87–92 (1976)
© 1976 by Dr. Dietrich Steinkopff Verlag GmbH & Co. KG, Darmstadt
ISSN 0340-255 X

Plenary lecture of the IUPAC-Conference on Colloid and Surface Science in
Budapest, September 15–20, 1975

*Akademie der Wissenschaften der DDR, Zentralinstitut für physikalische Chemie,
Abteilung Kolloidchemie, Berlin (DDR)*

Stern potential, zeta potential and dipole moment of aerosil particles dispersed in electrolyte solutions

H. Sonntag and *H. Pilgrimm*

With 3 figures

(Received December 9, 1975)

1. Electrochemical double layer potentials at the solid/liquid phase boundary

The charge of particles in water solutions results from the transfer of ions from the solid into the solution phase or vice versa or from the dissociation of polar groups at the solid liquid surface or from the adsorption of ions out of the solution phase.

In the simplest case the electrochemical double layer consists of charge carriers on the solid surface and an equivalent quantity of oppositely charged solvated charge carriers in the solution phase. Due to the electrostatic attractive forces between the charge on the solid surface and the counterions in the solution phase, a charge distribution results where a more or less high portion of the solvated counterions is arranged at the smallest possible distance from the phase boundary (outer Helmholtz layer). As a result of thermal motion, the remaining solvated counterions are diffusely distributed towards the interior of the solution phase, their density decreasing in this direction.

The equilibrium Galvani voltage $\Delta\varphi$, which results from the difference of the inner electrical potentials φ of the adjacent phases, consists of the contact voltage $\Delta\psi$, i.e. of the difference between the external potentials ψ of the adjacent phases, which is determined by the amount and the sign of the excess charge, and the difference between the so-called surface potentials $\Delta\chi$ determined by the amount and the orientation of oriented dipoles:

$$\Delta\varphi = \Delta\psi + \Delta\chi . \tag{1}$$

In the colloid chemistry especially the external potentials caused by the excess charges are of interest. Another reason for their special importance is that they represent the only measurable phase boundary potentials.

The external potential in the interior of the liquid phase is set equal to zero: $\psi_\infty = 0$, and the external potential at the surface of the solid phase is denoted by ψ_0. Thus one obtains from [1]:

$$\psi_0 = \Delta\varphi - \Delta\chi . \tag{2}$$

If there is no excess charge at the surface of the solid phase, then the Galvani voltage across the double layer becomes equal to the dipole voltage, since $\psi_0 = 0$. Hence at this point of zero charge one obtains:

$$\Delta\varphi_{\text{p.z.c.}} = \Delta\chi . \tag{3}$$

If it is assumed that the dipole voltage portion remains constant with the variation of the surface charge density, one obtains:

$$\psi_0 = \Delta\varphi - \Delta\varphi_{\text{p.z.c.}} . \tag{4}$$

However, this approximation holds only if the structure of the dipole layer does not change, which is practically possible only for surface potentials in the neighbourhood of the point of zero charge or for systems where the orientation of the dipole molecules is effected by hydrogen bonds at the solid surface (e.g. at non-porous oxide surfaces).

A problematic task is the determination of ψ_0 at oxide surfaces, since till now there are no electrodes available which permit measuring the ψ_0 potentials. The thermodynamic methods of calculating ψ_0 are likewise rather uncertain, since it is very difficult (if not impossible) to take into account dipole voltage variations.

The potentials at the outer Helmholtz layer ψ_δ and the electrokinetic potential ζ lying at the shear plane of the liquid phase, which is moving relative to the solid surface, are of special importance because they permit statements to be made on the stability of dispersions and the like. The ζ-potential is readily observable by experiment, although the individual measuring techniques frequently provide different results of measurement. The theoretically most certain results are obtained with the use of electrophoresis described by *Wiersema* and *Overbeek* (1).

The ψ_δ-potential is not directly measurable; it can only be derived from other measured quantities.

A possibility of determining the ψ_δ-potential is given in cases where the surface charge density at the solid surface can be experimentally determined and if there is no specific adsorption of ions in the inner Helmholtz layer (2).

In many cases, however, it is not possible to determine the surface charge density, so that the ψ_δ-potential cannot be determined. In the colloid chemical literature we find two other methods to determine the potential of the outer Helmholtz plane: 1. it is possible to calculate ψ_δ-potentials from model experiments on thin liquid films; foam films (3) and emulsion films (4), by measurement of the equilibrium thickness of these films under various conditions.

2. From flocculation experiments with particles of known radius *Reerink* (5) and *Ottewill* (6) estimated ψ_δ-potentials from the slope of the stability ratio-log c-curves. Both the methods are of low accuracy and require a high experimental effort.

As it shall be explained in more detail in the following section, there is however a possibility of determining the ψ_δ-potential from conductivity measurements if we have an independent method for measuring ζ-potentials with high accuracy.

2. Electric conductivity of dispersions of uncharged particles

If non-conducting particles are added to an electrolyte solution, then the electric conductivity k_s of the produced dispersion decreases according to the following relation:

$$k_s - k_0 = -4\pi N \cdot \mu\, \frac{k_0}{E} \qquad [5]$$

where

μ = dipole moment of the particles,
N = number of particles per unit volume of the disperse phase,
E = electric field strength,
k_0 = electric conductivity of the electrolyte solution.

If the volume part p of the disperse phase is introduced by $p = \frac{4}{3} a^3 \cdot N$, then [5] yields:

$$k_s - k_0 = -\frac{3p}{a^3}\, \mu \cdot \frac{k_0}{E}\,. \qquad [6]$$

If an electric field is applied to a colloidal dispersion with a conductive dispersion medium, then the cations migrate towards the cathode and the anions towards the anode. If the ions strike a particle on their way, then the cations will be trapped on the one side, the anions on the other side of the particle. The cations trapped on the one side of the particle will repel the new cations coming from the interior of the volume, and the anions on the other side of the particle will repel the anions coming from the solution. The more ions accumulate on the surface, the stronger the repulsion forces will be. The state of equilibrium is reached when the supply of ions from the interior of the solution towards the particle surfaces is finished, i.e. when the polarization field produced by the ions accumulating at the particle surfaces completely compensates the applied external electric field. Then the normal component E_n of the field, which is responsible for the migration of ions from the interior of the volume, is zero. Consequently also the migration of the ions towards the surface is zero, i.e. there is no surface conductivity at uncharged particles.

The potential developed at a spherical nonconductive particle under the influence of an external electric field is calculated (7) according to:

$$\varphi = -E\left(r + \frac{a^3}{2r^2}\right)\cos\vartheta \qquad [7]$$

where

r = distance from the particle centre,
a = particle radius.

The potential distribution differs from that in a homogeneous field

$$\varphi_\infty = -E \cdot r \cdot \cos\vartheta \qquad [8]$$

by the term

$$\varphi_1 = \frac{a^3}{2\,r^2} \cdot E \cdot \cos \vartheta \,. \qquad [9]$$

Since for the potential of the dipole having the dipole moment μ, the following relation holds for large distances

$$\varphi_\mu = -\frac{\mu}{r^2} \cos \vartheta \,. \qquad [10]$$

Comparing eqs. [9] and [10] one observes that the dipole moment of an uncharged particle in a conductive medium with an electric field applied can be calculated according to

$$\mu = \frac{a^3}{2} \cdot E \,. \qquad [11]$$

Thus by substituting eq. [11] into eq. [6] the variation of electric conductivity of a dispersant caused by adding uncharged particles can be calculated according to

$$k_s - k_0 = -\tfrac{3}{2}\, p \cdot k_0 \,. \qquad [12]$$

From this equation follows that the electric conductivity of the dispersion decreases with increasing volume portion of the disperse phase since the cross section of ion transport decreases.

3. Electric conductivity of dispersions containing charged particles

Under the influence of an external electric field, the electrochemical double layer of charged particles in a colloidal dispersion is deformed, so that the individual particles develop dipole moments with orientations opposite to the external field.

According to *Duchin-Semenichin* (8), the potential of the deformed double layer of charged particles can be calculated by the following relation:

$$\varphi = -\left[r + \left(\frac{1}{2} + A\right)\frac{a^3}{r^2}\right] E \cos \,. \qquad [13]$$

The constant A characterizes the polarization of the double layer and can be expressed, after rearranging and reducing the above relation, by

$$A = -\frac{3\,\mathrm{Rel}}{(2\,(1 + 2\,\mathrm{Rel}))} \qquad [14]$$

where the parameter Rel gives us a relation between ζ and ψ_δ-potentials:

$$\mathrm{Rel} = \frac{1}{\varkappa\, a}\left[4\,(1 + 3\,m) \sin h^2 \frac{\tilde{\zeta}}{4} \right.$$
$$\left. + 2\left(\cos h \frac{\tilde{\psi}_\delta}{2} - \cos h \frac{\tilde{\zeta}}{2}\right)\right] \qquad [15]$$

$\varkappa\ =$ Debye-Hückel parameter,
$m = \dfrac{(k\,T)^2\, \varepsilon\, \varepsilon_0}{e^2 \cdot 6\,\pi\,\eta\,D}$,
$\tilde{\psi}_\delta = \dfrac{e \cdot \psi_\delta}{k\,T}$,
$\tilde{\zeta}\ = \dfrac{e \cdot \zeta}{k\,T}$,
$k\ =$ Boltzmann constant,
$T\ =$ absolute temperature,
$\eta\ =$ dynamic viscosity of the dispersant,
$D\ =$ diffusion coefficient,
$e\ =$ elementary charge,
$\varepsilon\ =$ dielectric constant,
$\varepsilon_0 =$ influence constant.

Comparing eq. [13] with eq. [8], one obtains the following expression for the difference between the potentials of the deformed double layer and that in the homogeneous field:

$$\varphi_2 = -\left[\left(\frac{1}{2} - \frac{3\,\mathrm{Rel}}{2 + 4\,\mathrm{Rel}}\right)\frac{a^3}{r^2}\right] E \cos \vartheta \,. \qquad [16]$$

Comparing the dipole potentials according to eq. [10] with that of eq. [16], one observes that the dipole moment of a charged particle in an external electric field can be calculated by

$$\mu = \frac{a^3}{2} \cdot E\left(1 - \frac{3\,\mathrm{Rel}}{1 + 2\,\mathrm{Rel}}\right). \qquad [17]$$

The electric conductivity of dispersions containing charged particles can be calculated by substituting eq. [17] into eq. [6] according to

$$k_s = k_0 - \frac{3}{2}\, p\left(1 - \frac{3\,\mathrm{Rel}}{1 + 2\,\mathrm{Rel}}\right) k_0 \,. \qquad [18]$$

Thus it is possible to calculate Rel by conductivity measurement of the dispersion and the dispersion medium. Then the ψ_r-potential can be calculated by determining the ζ-potential.

As Rel increases ($\mathrm{Rel} \approx 1$), the term in brackets in eq. [18] may become zero, i.e.

$$\frac{3\,\mathrm{Rel}}{1 + 2\,\mathrm{Rel}} = 1 \,.$$

Subject to this assumption, $k_0 = k_s$.

This value of electric conductivity, at which the conductivity of the disperse system is equal to the equilibrium conductivity of the electrolyte solution, is called the iso-conductivity. Here the conductivity-reducing effect of the low-conducting particles is compensated by the surface conductivity in the diffuse double layer of these particles. The surface conductivity k^σ depends on the structure of the solid/electrolyte solution phase boundary, especially on the surface charge density and the mobility of ions in the diffuse double layer. If one assumes that the charge in the diffuse part of the double layer is determined mainly by the counter-ions, then due to the electric neutrality of the double layer the surface charge density of the disperse particles must correspond to the charge number σ_d in the diffuse part of the double layer (no specific adsorption). As the surface charge density increases, σ_d increases, and hence also k^σ. If the concentration c of the indifferent electrolytes is increased, then σ_0 and hence also σ_d, remain constant. Only the thickness of the diffuse double layer will decrease. Consequently the surface conductivity does not depend on c but only on σ_0 and σ_d. In contrast to this, the electrolyte conductivity strongly depends on c, and hence also on Rel. Therefore also the dipole moment of charged particles is strongly influenced by c.

4. Experimental determination of the surface charge density

As a system to be investigated we used dispersions of Aerosil OX 50 in aqueous KCl solutions. Aerosil is a flam hydrolysed silica having silanol groupings at its surface which can reversibly be converted into silanolate ions by reaction with hydroxyl ions in aqueous dispersions.

$$Si{-}OH + OH^- \rightleftharpoons Si{-}O^- + H_2O \,.$$

Thus the number of the silanolate ions produced, and hence the surface charge density of the particles, can be varied within a wide range by variation of the hydroxyl ion concentration in the dispersion. Since the surface charge density depends, in addition to the hydroxyl ion concentration, also on the concentration of counter-ions, it was not possible to apply a potentiometric titration as it was described in the literature for similar investigations of oxidic particles (9, 10). In the investigations cited, the adjustment of a certain pH value in the dispersions with given concentrations of counter-ions is achieved by the addition of alkali. This increases, however, the concentration of counter-ions during the titra-

tion. This results in an additional increase of the surface charge density, particularly if low counter-ion concentrations are involved ($c = 10^{-2}$ M).

By means of chloride-ion selective electrodes, we were able to find that the negative adsorption of chloride ions is negligible up to pH < 10, so that the charge density of the double layer is not influenced by chloride ions. Consequently the pH value of the dispersion can be varied by means of a variation, say, of the HCl content with a given constant KOH concentration and without varying the concentration of potassium ions.

The pH measurement was carried out under N_2-flushing at 25 °C. From the difference between the amount of hydroxyl ions added to the dispersion and the concentration of hydroxyl ions measured in the electrolyte solution, it is possible to calculate the number of silanolate ions produced, and if the specific surface of the dispersed Aerosil particles is known, the surface charge density can be calculated. The surface charge density is defined by

$$\sigma_0 = F\,(\Gamma_{H^+} - \Gamma_{OH^-})\,,$$

where Γ_{H^+} and Γ_{OH^-} mean the quantities of H$^+$ and OH$^-$, respectively, per cm^2 of surface area, which are adsorbed or converted at the phase boundary, and $F =$ Faraday constant.

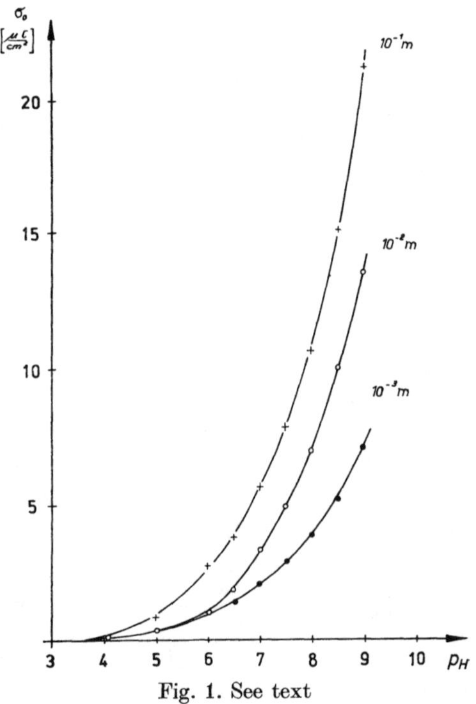

Fig. 1. See text

In fig. 1, the σ_0 vs. pH curves are plotted for different concentrations of potassium ions.

From the surface charge densities determined in this way, it is possible to calculate the corresponding potentials ψ_δ at the outer Helmholtz layer (2).

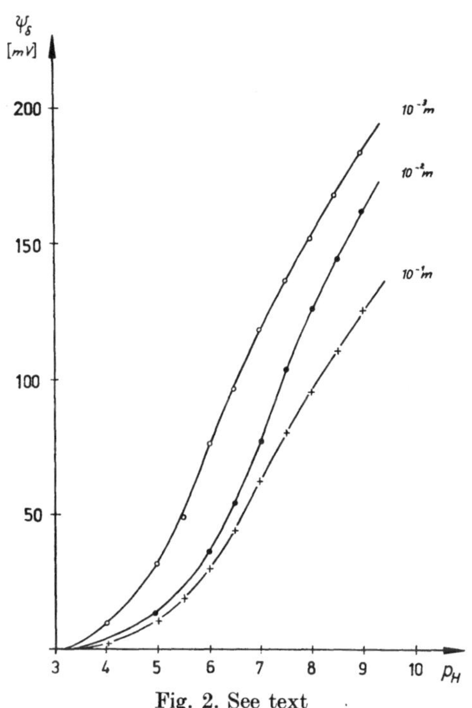

Fig. 2. See text

In fig. 2, the ψ_δ vs. pH curves are plotted for various KCl solutions.

5. Experimental determination of electric conductivity

The investigation of the electrical conductivities of the dispersion and the ultra-centrifugates was carried out together with the determination of the surface charge density in one and the same measuring cell.

For the determination of electric conductivity we used an impedance bridge. The measuring frequency was adjusted to 2 kHz in order to provide sufficient time for the reorientation of the induced dipole moments of the particles. According to *Einstein*, a time $\tau = b^2/D$ is required for a concentration distribution to develop within a region with linear dimension b. With $b \sim 10^{-6}$ cm and $D \sim 10^{-5}$ cm^2 s^{-1}, one obtains a reorientation time of $\tau \sim 10^{-7}$ s. Hence for the measuring frequency used the reorienta-

tion of the dipole moments in the applied external alternating field is ensured.

Lower frequencies should be avoided because of the occurrence of electrode polarization.

In order to determine the dispersion conductivity k_s and the corresponding electrolyte conductivity k_0, we divided the dispersions into two portions. The one half we used for investigating the dispersion conductivity, in the other half we separated the Aerosil from the electrolyte solution by ultracentrifugating and determined the conductivity and the pH-value of the electrolyte solution.

6. Discussion of the results

As it is observed from the σ_0 vs. pH curves shown by fig. 1, the surface charge density of Aerosil particles increases with increasing concentration of the potassium ions and constant pH value. The point of zero charge lies at the pH of 3.5, i.e. at this pH all of the silanol groups are in the undissociated state, the surface charge density and the surface potential ψ_0 both being zero. Consequently in this case we have a dispersion of uncharged particles in an electrically conductive dispersion medium. For such systems, eq. [12] holds for the electric conductivity:

$$k_s = k_0 \tfrac{3}{2} p \, k_0 .$$

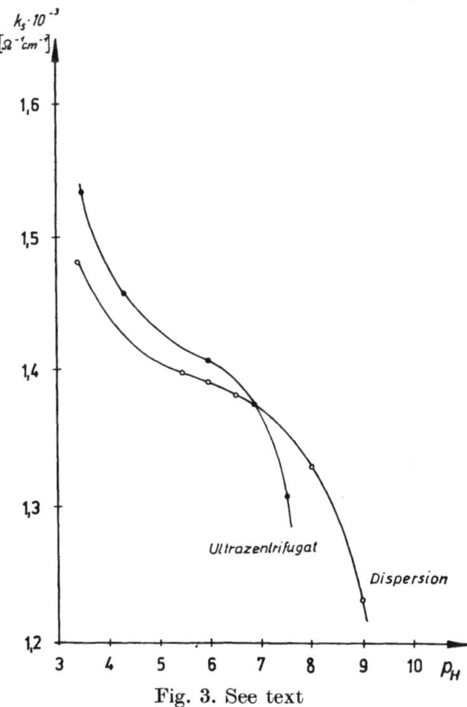

Fig. 3. See text

Hence for the 5% dispersion considered in fig. 1, a relative conductivity reduction $(k_s - k_0)/k_0$ of -0.023 results. From fig. 3, a relative conductivity reduction of -0.025 is calculated for a pH of 3.5. The occurring difference between the calculated reduction of conductivity and that determined from conductivity measurements is within the limits of error of the measuring methods.

As it is observed from fig. 1, the surface charge density of the particles increases with increasing pH. Simultaneously the surface conductivity and Rel increase.

If Rel increases (Rel $\approx$ 1), the term in brackets in eq. [18] may become zero. Subject to this condition, $k_s = k_0$, i.e. the electric conductivity of the dispersion is equal to that of the dispersion medium. This point of iso-conductivity is found at the point of intersection of the conductivity curves of the dispersion and the ultra-centrifugate in fig. 3.

As the pH further increases, the surface charge density of the particles, and hence also their surface conductivity will increase. The term Rel will become greater than 1. Consequently the conductivity of the dispersion will exceed that of the electrolyte solution.

If the volume part of the disperse phase is known, there can be determined by measuring the conductivities of the dispersion and the ultra-centrifugate Rel. The knowledge of Rel now permits calculating the potential ψ_δ at the outer Helmholtz layer, if the ζ-potential has been determined by experiment.

The determination of ζ-potentials was performed using a mass transport cell (11), the accuracy and reproducibility of this method being rather unsatisfactory.

Till now a quantitative check of the ψ_δ determination from conductivity measurements did not provide a very good agreement between the ψ_δ-potentials calculated from the determined surface charge densities and the ψ_δ-potentials determined from conductivity measurements and ζ-potentials.

Thus from measurements of surface charge density in 2,5% Aerosil OX 50 dispersions at KCl concentrations of, say, 10^{-2} M, a ψ_δ-potential of 78 mV and a ζ-potential of 50 mV was measured at the iso-conductivity point. However, from conductivity measurements and the ζ-potential we calculated a ψ_δ-potential of 92 mV.

At present the development of a semi-automatic measuring apparatus for ζ-potentials is under way, so that we hope that it is possible in the near future to perform a more accurate verification of the theory.

References

1) *Wiersema, P. H.*, Dissertation (Utrecht 1964).
2) *Loeb, A., P. H. Wiersema*, and *J. Th. G. Overbeek*, J. Colloid Sci. **22**, 78 (1966).
3) *Exerowa, D.*, Kolloid-Z. u. Z. Polymere **232**, 703 (1969).
4) *Sonntag, H., J. Netzel*, and *B. Unterberger*, Disc. Faraday Soc. **42**, 57 (1970).
5) *Reerink, H.* and *J. Th. G. Overbeek*, Disc. Faraday Soc. **18**, 74 (1954).
6) *Ottewill, R. H.* and *J. N. Shaw*, Disc. Faraday Soc. **42**, 154 (1966).
7) *Overbeek, J. Th. G.*, Kolloid-Beihefte **54**, 287 (1943).
8) *Duchin, S. S.* and *N. M. Semenichin*, Kolloid-J. (russ.) **32**, 360 (1970).
9) *Bolt, G. H.*, J. Phys. Chem. **61**, 1166 (1957).
10) *Tadros, Th. F.* and *J. Lyklema*, J. Electroanal. Chem. **17**, 267 (1968).
11) *Long, R. P.* and *S. Ross*, J. Colloid Sci. **20**, 438 (1965).

Authors' address:

H. Sonntag and *H. Pilgrimm*
Akademie der Wissenschaften der DDR
Zentralinstitut für physikalische Chemie
Abteilung Kolloidchemie
Rudower Chaussee 5
DDR-1199 Berlin-Adlershof

Progr. Colloid & Polymer Sci. **61**, 93–96 (1976)
© 1976 by Dr. Dietrich Steinkopff Verlag GmbH & Co. KG, Darmstadt
ISSN 0340-255 X

Plenary lecture of the IUPAC-Conference on Colloid and Surface Science in
Budapest, September 15–20, 1975

Chemical Research Institute of Non-aqueous Solution, Tohoku University, Katahira, Sendai (Japan)

Surface energy analysis of solids and its application

Y. Tamai

With 2 figures and 4 tables

(Received December 9, 1975)

Introduction

Since 1964 when *Fowkes* has proposed the method to analyse surface tension, γ, into its components such as dispersion force term, γ^d, electrostatic force term, γ^e, and so forth (1), many studies have been conducted to extend Fowkes' method and to apply the results for adhesion and some other surface phenomena. This report belongs to the same category of research and the aim is first to explain the theoretical base of Fowkes' expansion and to compare the two contact angle methods of analysis which leads to the estimation of surface pressure of the employed liquid. The latter part of this report is concerned with the adhesion or friction of polymer plastics, and the experimental results are shown to have good correlation with the calculated work of adhesion.

Theoretical base of surface tension expansion

The expansion of surface tension is expressed as

$$\gamma = \gamma^d + \gamma^e + \cdots . \tag{1}$$

However, the surface tension is surface excess free energy and depends not only on energy term but also on entropy term. The energy term may be expanded due to intermolecular forces. As for the entropy term, no direct evidence or proof has been submitted, and indeed some critics were discussed on this problem (2).

Recently a simple cell model treatment of this expansion was proposed (3). Applying Lennard-Jones' liquid model to the molecular expression of surface tension,

$$\gamma = \frac{1}{\sigma}\left[\frac{1}{2}(W_s - W_b) + kT \ln\left(\frac{v_b}{v_s}\right) \right], \tag{2}$$

where σ is the area of one surface molecule, W_s and W_b the energy of the surface and the bulk molecule, respectively, and v_b and v_s the vibration frequency of the bulk molecule and the perpendicular component of that of the surface molecule. T is the absolute temperature and k the Boltzmann constant.

W can be expanded into dispersion W^d, electrostatic W^e, and so forth. The problem is whether or not the entropy term of $\ln(v_b/v_s)$ can be also expanded into additive series due to intermolecular forces. v can be expressed by force constant K and molecular mass m as

$$v = \frac{1}{2\pi}\left(\frac{K}{m}\right)^{1/2} . \tag{3}$$

Therefore so far as the K_s, that of perpendicular vibration to the surface, differs by small amount ΔK from the K_b, in the bulk,

$$\ln\left(\frac{v_b}{v_s}\right) = \frac{1}{2}\ln\left(\frac{K_b}{K_s}\right) = \frac{1}{2}\ln\left(\frac{K_b}{K_b - \Delta K}\right)$$
$$= -\frac{1}{2}\ln\left(1 - \frac{K}{K_b}\right) = \frac{1}{2}\frac{\Delta K}{K_b} . \tag{4}$$

It is easily seen that the force constant K can be additively expanded into dispersion K^d, electrostatic K^e, and so forth. Then, after all, γ itself can be expanded into γ^d, γ^e, and so forth, if the harmonic vibration is approximately assumed with the molecule or the segment of molecule in the bulk and in the surface.

The two analytical methods and the surface pressure

The simplest way of analysis is to measure the contact angle, θ, of a liquid on the plain surface of the given solid. Young-Dupré relation gives

$$\gamma_S = \gamma_{SL} + \gamma_L \cos\theta , \tag{5}$$

W 406

where γ_S and γ_L are the surface tension of solid and liquid, respectively, and γ_{SL} the interfacial tension of the solid and the liquid. The interfacial tension can be regarded as

$$\gamma_{SL} = \gamma_S + \gamma_L - I_{SL} \,. \qquad [6]$$

I_{SL}, the work of adhesion, is further expressed as

$$\begin{aligned} I_{SL} &= I_{SL}^{\mathrm{d}} + I_{SL}^{\mathrm{e}} + \cdots \\ &= 2(\gamma_S^{\mathrm{d}} \gamma_L^{\mathrm{d}})^{1/2} + 2(\gamma_S^{\mathrm{e}} \gamma_L^{\mathrm{e}})^{1/2} + \cdots \,. \end{aligned} \qquad [7]$$

This expression may be open to discussion. *Fowkes* showed its theoretical derivation (1), and as for I_{SL}^{e}, *Hata* proposed it just as an assumption (4). These expressions are commonly regarded as Berthelot's relation. As shown later, the relation [7] holds experimentally to considerable extent.

Combining the relations [5], [6], and [7],

$$\begin{aligned} \gamma_L(1 &+ \cos\theta) \\ &= 2(\gamma_S^{\mathrm{d}} \gamma_L^{\mathrm{d}})^{1/2} + 2(\gamma_S^{\mathrm{e}} \gamma_L^{\mathrm{e}})^{1/2} + \cdots \,. \end{aligned} \qquad [8]$$

If a saturated hydrocarbon liquid H is employed, $\gamma_H^{\mathrm{d}} = \gamma_H$, and any other term than dispersion force can be neglected.

$$\gamma_H(1 + \cos\theta) = 2(\gamma_S^{\mathrm{d}} \gamma_H)^{1/2} \,. \qquad [9]$$

If a polar liquid, for example water W, which is known of γ^{d} 23.0 and γ^{n}, non-dispersion force component, 49.7, is employed (5),

$$\gamma_W(1 + \cos\theta) = 2(\gamma_S^{\mathrm{d}} \gamma_W^{\mathrm{d}})^{1/2} + 2(\gamma_S^{\mathrm{n}} \gamma_W^{\mathrm{n}})^{1/2} \,. \qquad [10]$$

Therefore, γ_S^{d} or γ_S^{n} can be obtained experimentally by obtaining θ for one more liquid such as glycerine.

This kind of method has its restriction. The solid must not be spreadingly wettable with the employed liquid. In this method the contact angle is measured against air or gas, and in this sense this method can be specified as the one-liquid method. To obtain contact angle with high energy solid, a system of two liquids and one solid, for example water, hydrocarbon liquid, and metal, may be satisfactorily employed (6). For this system the fundamental relation is

$$\begin{aligned} \gamma_H &- 2(\gamma_S^{\mathrm{d}} \gamma_H)^{1/2} \\ &= \gamma_W - 2(\gamma_S^{\mathrm{d}} \gamma_W^{\mathrm{d}})^{1/2} - I_{SW}^{\mathrm{n}} + \gamma_{SW} \cos\theta \,. \end{aligned} \qquad [11]$$

This equation contains two unknowns, that is, γ_S^{d} and I_{SW}^{n}. To solve this, another liquid hydrocarbon H' may be employed to get another relation of eq. [11]. This method is specified as the two-liquid method.

To compare the one-liquid and the two-liquid methods experimentally, the system of low energy solid, that is, polytetrafluoroethylene (PTFE), polyvinylchloride (PVC), and polymethylmethacrylate (PMMA), and almost non-polar liquid, that is, methylene iodide(met-I) and α-bromonaphthalene (Br-naph) were employed for the one-liquid method. The same polymer and water and hydrocarbon liquid, that is, n-heptane (n-hep), n-hexane (n-hex), iso-octane (i-oct), and cyclohecane (c-hex) were used for the two-liquid method. The experimental details will be reported elsewhere. The obtained γ_S^{d}'s are listed in table 1.

Table 1. γ_s^{d} of polymer solid (erg/cm^2, 20 °C)

Method	Liquid	PTFE	PVC	PMMA
one-liquid	met-I	16	39	42
	Br-naph	17	42	44
	(average)	(17)	(41)	(43)
two-liquid	c-hex/n-hep	25	61	96
	c-hex/n-hex	25	58	90
	c-hex/i-oct	25	56	88
	(average)	(25)	(58)	(91)

γn: met-I 50.8, Br-naph 44.5, c-hex 24.9, n-hep 20.2, n-hex 18.7, i-oct 18.9.

It is obvious that there is definite difference. To explain this difference, the nature of Young-Dupré equation was examined. In eq. [5], γ_S is in its left side, but exactly it should be γ_{SV} which is the surface tension of solid covered with adsorbed vapor, that is,

$$\gamma_{SV} = \gamma_S - \pi_V \,, \qquad [12]$$

where π_V is the surface pressure of that liquid. Then eq. [9] should be

$$\gamma_N(1 + \cos\theta) = 2(\gamma_S^{\mathrm{d}} \gamma_N)^{1/2} - \pi_N \,. \qquad [13]$$

Subscript N represents non-polar liquid. Sometimes π_N is neglected, because π_N is assumed to be very small if $\gamma_N > \gamma_S$ (7). However, there are also several experimental studies which shows considerable π value. For example, π of hexane on PTFE is reported 5.7 dyne/cm from adsorp-

Table 2. π_N on polymer solids (dyne/cm, 20 °C)

vapor	PTFE	PVC	PMMA
methylene iodide	13	19	44
α-bromonaphthalene	9	15	40

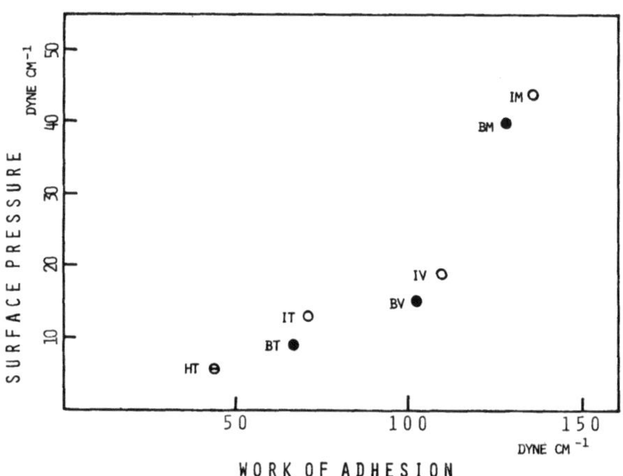

Fig. 1. Surface pressure and work of adhesion. (H: n-hexane, B: bromonaphthalene, I: methylene iodide, T: PTFE, V: PVC, M: PMMA, vapor and solid, in order)

tion measurement (8). Therefore, it may be interesting to put γ_S^{d} values from the two-liquid method into eq. [13] to get π_N's. Table 2 shows the results. In fig. 1 these π_N's are plotted against I_{SL}^{d} for each liquid together with the literature value of hexane on PTFE. The tendency that the higher π corresponds to the larger I_{SL}^{d} seems reasonable.

The two-liquid method gives both γ_S^{d} and I_{SW}^{n}. In this case, eq. [13] should be modified to

$$\gamma_W(1 + \cos\theta) = 2(\gamma_S^{\mathrm{d}}\,\gamma_W^{\mathrm{d}})^{1/2} + I_{SW}^{\mathrm{n}} - \pi_W . \quad [14]$$

From this equation, π_W was estimated as 3, 5 and 23 dyne/cm (20 °C) on PTFE, PVC, and PMMA, respectively. π_W on graphon ($\gamma_S^{\mathrm{d}} = 70$) was reported as 6 (8), which is comparable with the present data.

The analysed component of surface tension and the static friction

As the two-liquid method gives I_{SW}^{n}, γ_S^{n} is calculated by putting γ_W^{n} as the difference between γ_W and γ_W^{d}, and applying it into the so-called Berthelot's relation. In table 3 the data are summarized for organic polymer solids, in which γ_W^{n} is taken as 50 dyne/cm. With PTFE and mylar, some data have been reported both γ_S^{d} and γ_S^{n} (5). These data were obtained by the one-liquid (polar) method employing water and glycerine ($\gamma_S^{\mathrm{d}} = 34$, $\gamma_S^{\mathrm{n}} = 30$), without any consideration of the surface pressure of the liquid. Nevertheless, as is shown in table 4, mutual agreement is rather good, probably due to the small or negligible surface pressure of

Table 3. γ_S^{d} and γ_S^{n} of several polymers (erg/cm², 20 °C)

polymer	γ_S^{d}	γ_S^{n}
polyethylene	24 ± 3	0.01
polystyrene	46 ± 1	0.05 ± 0.02
polytetrafluoroethylene	25	0
polyvinylchloride	58 ± 3	0.8 ± 0.1
polymethylmethacrylate	91 ± 5	3.1 ± 0.1
mylar	50 ± 2	4.4 ± 0.1
neoprene	31 ± 9	0.02 ± 0.01
phenol resin	74 ± 5	5.1 ± 0.1
epoxy resin	54 ± 8	3.0 ± 0.1
melamine resin	82 ± 12	10.5 ± 0.2

Table 4. Comparison with other method

polymer	γ_S^{d}	γ_S^{n}	method (source)
PTFE	25	0	two-liquid (*Tamai* et al.)
	20 ± 4	0.4 ± 0.8	one-liquid (*Andrews* et al.)
mylar	50 ± 2	4.4 ± 0.1	two-liquid (*Tamai* et al.)
	42 ± 7	3.3 ± 2.8	one-liquid (*Andrews* et al.)

water and glycerine on the examined polymers. This agreement also suggests that the Berthelot's relation holds at least experimentally.

It is a very interesting problem, especially from the standpoint of industrial application, to investigate whether these data on γ^{d} and γ^{n} can be useful to estimate wetting property or adhesive strength of given engineering materials. For this purpose, static friction was measured with a sliding couple of polymers. The static friction can be regarded as measure of shear adhesive strength. The mechanism of friction of plastic is complicated and the interpretation of the results is not so easy. However, for the first approximation, the coefficient of static friction is representative of adhesion, if measured under a sufficiently light load to avoid any larger amount of deformation at the real contact.

Some results are illustrated in fig. 2, in which the work of adhesion calculated from eq. [7] for each couple is on abscissa. It is clear from this figure that the friction couple of larger work of adhesion exhibits higher friction coefficient. In connection with the experiment, the effect of humidity must be taken into consideration. In the humid atmosphere the friction was observed

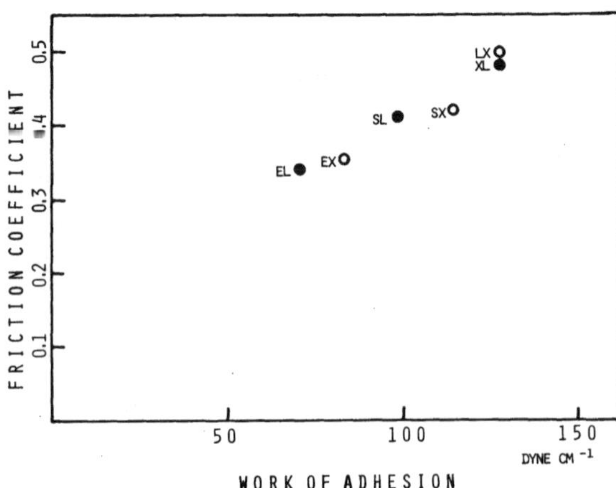

Fig. 2. Static friction and work of adhesion (load: 0.2 g, dry atm., E: polyethylene, S: polystyrene, L: epoxy resin, X: phenol resin, base and slider, in order)

higher than in dry atmosphere. This humidity effect was more evident with the couple of larger non-dispersion force term of the adhesion work. For the sliding under very light load, adsorbed water seems to behave as a kind of adhesive. Further quantitative research may be desirable on this phenomenon.

Further remarks and problems on the analysis

Methods to characterize solid surface have been developed recently, for example, ESCA or Auger electron spectroscopy. However, information obtained by some of these methods is restricted, because of their special measuring conditions such as under ultra-high vacuum. The information of solid surface is very important under industrial or common conditions. In this aspect, surface tension analysis by contact angle or adsorption measurements is of high potential applicability. Surface tension itself is a macroscopic or phenomenological quantity, but the expansion of it into its components due to the nature of intermolecular forces gives microscopic or atomistic understanding based on molecular characteristics, which is valuable to design engineering materials of desired surface properties.

There are some problems remained to be solved in the surface tension analysis. First is to establish the method to obtain reliable and reproducible results. It depends on how to prepare a standard solid surface and how to measure contact angle as accurate as possible. For the two-liquid method, it is recommended to employ hydrocarbon liquids of widely different surface tension.

Another problem is fundamental on the nature of contact angle measured. Advancing contact angle is regarded as in equilibrium (9). However, if any physical or chemical change is expected at the interface of the solid and the liquid, receding angle may differ from advancing, and in this case the advancing angle may not be regarded of thermodynamic meaning. With the one-liquid method, as pointed out above, the neglect of the surface pressure is under question. With the two-liquid method, the complete substitution of matrix liquid with dropping or immersing liquid must be assured at the interface.

Summary

The expansion of surface tension into an additive series of terms due to the intermolecular forces, originally introduced by *Fowkes*, is firstly discussed. Based on this expansion, a survey is given on the analytical methods to determine the terms by means of contact angle measurements. An examination on these methods and some results on several organic polymer solids is made, and the surface pressure of adsorbed vapor is estimated. Also assuming the Berthelot's relation on the interfacial energy, a polar term of surface tension is derived from the experimental data, in addition to the non-polar term. As an application of this analysis, the calculated interfacial energy is compared with the shearing adhesive strength between polymeric solids, which is approximated to the static friction.

References

1) *Fowkes, F. M.*, Ind. Eng. Chem. **56**, 40 (1964).
2) *Johnson, R. E.* and *R. H. Dettre*, Surface and Colloid Sci., vol. 2, p. 120 (New York 1960); *Hoernschemeyer, D.*, J. Phys. Chem. **70**, 2629 (1966); *Good, R. J.* and *E. Elbing*, Ind. Eng. Chem. **62**, 61 (1970).
3) *Tamai, Y.*, J. Phys. Chem. **79**, 965 (1975).
4) *Hata, T.*, Kobunshi **17**, 594 (1968).
5) *Andrews, E. H.* and *A. J. Kinloch*, Proc. Roy. Soc. **A 332**, 385 (1973).
6) *Tamai, Y., K. Makuuchi*, and *M. Suzuki*, J. Phys. Chem. **71**, 4176 (1967).
7) *Zisman, W. A.* and *I. P. Weiss*, Adhesion and Cohesion, p. 176 (New York 1962).
8) *Zettlemoyer, A. C.*, J. Colloid Interface Sci. **28**, 343 (1968).
9) *Gaudin, A. M.* and *A. F. Witt*, Contact Angle, Wettability, and Adhesion, Adv. Chem. Ser 43, Amer. Chem. Soc., p. 202 (Washington 1964).

Author's address:

Yasukatsu Tamai
Chemical Research Institute of Non-aqueous Solution
Tohoku University Katahira, Sendai 980 (Japan)

Progr. Colloid & Polymer Sci. **61**, 97–108 (1976)
© 1976 by Dr. Dietrich Steinkopff Verlag GmbH & Co. KG, Darmstadt
ISSN 0340-255 X

Plenary lecture of the IUPAC-Conference on Colloid and Surface Science in
Budapest, September 15–20, 1975

Equipe de Recherche C.N.R.S. associée á l'Université Paris V, U.E.R. Biomédicales, Paris (France)

Surface chemistry and wettability of modified polyethylene

*A. Baszkin, M. Deyme, M. Nishino**), and *L. Ter-Minassian-Saraga*

With 12 figures and 4 tables

(Received December 9, 1975)

I. Introduction

To modify the adhesivity of low energy polymers, surface and bulk treatments have been attempted. Among these treatments are the surface oxidation and bulk or surface grafting of monomers or polymers.

For polyethylene (PE), either limited (1) or intensive oxidations have been performed as well as microscopic grafting of polar constituents on linear macromolecules in solution (2) or on the surface of macroscopic samples of the polymer. The treated samples display better wettabilities than the original ones.

The improvement of wettability is one of the reasons of the increase in strength of adhesion (3).

Fowkes (4) and *Good* and *Girifalco* (5) have demonstrated the determinative effect of molecular interactions at the solid-liquid interface on the wettability of the polymer. Their interpretation may be applied to polymers which surface chemical composition is homogeneous.

In the case of modified surface these interpretations may be attempted when the modified polymer surface has been characterized.

Recently, methods based on the use of radioactive compounds (6) have been developed. They provide information about the chemical nature and the amount of the new surface groups produced by the chemical treatment of the polymers (6, 7, 10).

When these results are compared to those of wettability studies (8) it is found that, while the chemical modification progresses continuously, the wettability of the polymer reaches a limit.

This "saturation" effect cannot be understood unless it is assumed that the additional polar contribution to wettability is determined by the short range molecular interactions between the liquid and the added groups on the surface of the polymer.

These short-range interactions constitute the subject of our studies.

Such studies may be attempted on mixed polar-nonpolar polymers; e.g. compare polyoxyethylene (POE) with polyethylene (PE) and polyoxypropylene (POP). The critical surface tensions of these polymers are equal to 31 dyne/cm (PE); 32 dyne/cm (POP); 43 dyne/cm (POE) (11). POP critical surface tension is practically equal to that of the non-polar PE. The interpretation of this result is difficult as the three polymers have different long range dispersion forces, related to their bulk chemical composition, as well as different short-range interactions dependent on their surface chemical composition. Finally molecular orientation at the polymer surface may involve surface chemical compositions different from those in bulk.

The advantage of our approach is that the forces originating in the bulk of the polymer are not modified by the surface chemical treatment. The nature and the amount of the groups supplied to the surface of the polymer are varied and determined. Therefore we can separate quantitatively the contribution to wettability of short range interactions relevant to known polar groups on polymer surfaces.

To modify the surface of PE, it was either oxidized or grafted with poly-maleic acid (PMA). The oxidation was carried out by $KClO_3 + H_2SO_4$ mixtures (6, 7). Grafting was performed in a solution of maleic anhydride in acetic anhydride using benzoyl peroxide as initiator (7).

---

*) On leave from Institute of Chemical Research, Kyoto University (Japan).

Isotherms of Ca^{2+} adsorption on these surfaces were established at various pH and Na^+ concentrations for the grafted PE (7, 8) and in H_2O for the oxidized PE (6, 7). Parallel measurements of the contact angle of liquid drops placed on the characterized surfaces were performed using various organic liquids for the study of the oxidized samples (6, 7, 9) and aqueous salt solutions at various pH values for the grafted samples (10). Thus specific effects of counterions bound by the grafted PMA on the wettability of the modified PE have been found (10). Also, it has been shown that the relation of *Rehbinder* (12) corresponds well enough to a model interface between two quasi-cristalline condensed phases: the liquid and the oxidized polymer (9). The adhesion at this interface originates in the contacts between the surface molecules, if only next neighbor interactions exist (13). This model is correct for contacts between polar groups, as the carbonyls are formed by oxidation of PE (14).

The short-range action of dipolar liquid-solid interactions at interfaces is demonstrated also by studies of the effect of temperature on the adhesion tension of surface modified polymers (15, 16).

When chemical reactions take place at interfaces, covalent bonds are established between the solid and the liquid. A question has been raised by one of us (17): are *all* the surface molecules of the solid reacting with the liquid or is there an equilibrium established between the covalently bound surface molecules of the solid and the other surface molecules? In the second case, is it possible to relate the free energy of the surface reaction and the interfacial solid-liquid tension?

In the present paper we shall give a short description of the experimental techniques of the results and of the interpretation.

II. Experimental techniques

1. Preparation of the samples

a) Oxidation (6, 7)

Low density PE (0.929 g/cm³), 19 µ films were purified and oxidized at constant temperature, as described elsewere (6, 7) in mixtures of $KClO_3 + H_2SO_4$. The amount of oxidation increased with the concentration of $KClO_3$ in the mixture, but was independent of the duration of the attack. The standard duration of attack adopted was 0.5 minutes to 1.5 minutes.

The possible etching of the samples was too small to be detected by weighing.

b) Grafting on PE surface

Free radical polymerisation of maleic anhydride in acetic anhydride with benzoyl peroxyde acting as initiator was performed at 90 °C. An initial inhibition time of the reaction of 2 hours was noted. After extraction with acetone, the poly(maleic anhydride) was hydrolysed to PMA. The last step was controlled by IR absorption spectra of the samples. Details on this experiment may be found in reference (7).

The samples, grafted with PMA, have been changed to samples with grafted basic groups. The reaction is shown in table 1 and described in reference (18). It is controlled by IR spectroscopy. The final samples adsorb $S^{14}CN^-$ ions instead of $^{45}Ca^{2+}$ ions.

2. Contact angle measurements

The contact angles were measured on the samples analysed by radioactive ion adsorption. The apparatus is a temperature controlled goniometer operating below 120 °C. The drop on plate technique was used. Its accuracy was 2°. The heating rate of this apparatus was 0.25 °C/min and its temperature was known with an accuracy of ± 1 °C. The surface tension of the liquids was measured by the Wilhelmy plate method with a Pt blade, with an accuracy of 0.2 dyne/cm.

The melting of the PE samples was studied by differential thermal analysis with a conventional device.

3. Chemical characterization of the modified PE surface

The carbonyl groups formed by oxidation of PE adsorbed $^{45}Ca^{2+}$ ions from aqueous solutions. Based on this process, a technique for measuring the surface density of the carbonyls has been set up and described in reference (7).

Fig. 1 shows the apparatus which allows the measurement in situ of $^{45}Ca^{2+}$ ion adsorption at the solid-liquid interface.

The amount of grafted PMA was determined by the same technique. The basical grafted groups were determined by the adsorption of radioactive SCN (18) ions.

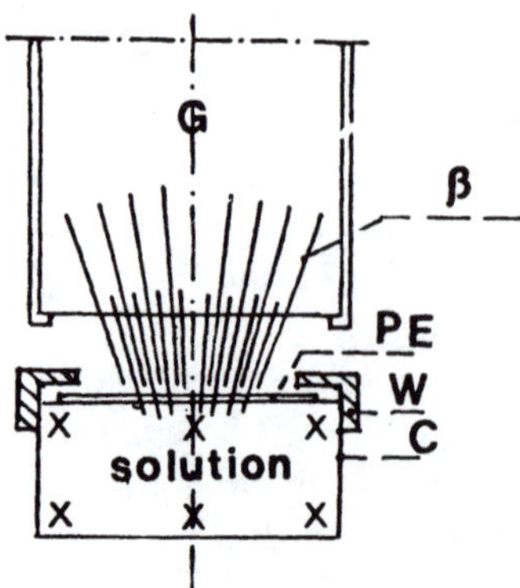

Fig. 1. Adsorption measuring apparatus; a: Geiger-Müller tube; PE: polyethylene film. W: teflon window; C: crystallizing dish. X: radioactive counterion

III. Results

1. Surface chemical characterization

a) Oxidized PE

The oxidation process was reproducible:

A given oxidizing mixture provided samples which adsorbed increasing amounts of Ca^{2+} ions when their concentration in solution increased.

The isotherms of adsorption were perfect Langmuir type I (7) with constants equal to 1.65×10^5 liters/mole^{-1}, and were independent of the surface density of the polar groups.

The maximum Ca^{2+} adsorption on a given sample was assumed to be equal to the surface density of the polar groups. On this basis, the curve 2 of fig. 2 has been drawn. It is noted that the increase of the $KClO_3$ concentration in the oxidizing mixture involves the parallel decrease in the contact angle.

Therefore the present method of oxidation may produce PE of controlled wettability.

Furthermore the samples are free of contamination of the heavy, chromic ions.

b) Grafted PE

Electron micrographs showed that the initial PE surface was uniformly smooth (ref. (19), fig. 3/2). Grafting introduced surface roughness and was not as reproducible as the oxidation process. A Gaussian distribution law has been obtained (19) when the number n_τ of samples displaying a given amount of grafting was plotted as a function of τ the amount of grafting (fig. 3). This result may be explained by the statistical distribution of the molecular weights of the grafted PMA chains.

The amount of grafting on each sample has been deduced from the adsorption isotherms of Ca^{2+} ions, assuming that at pH 12, the maximum adsorption of Ca ions corresponded to the saturation of the surface carboxyls, one Ca ion being bound to two carboxyls.

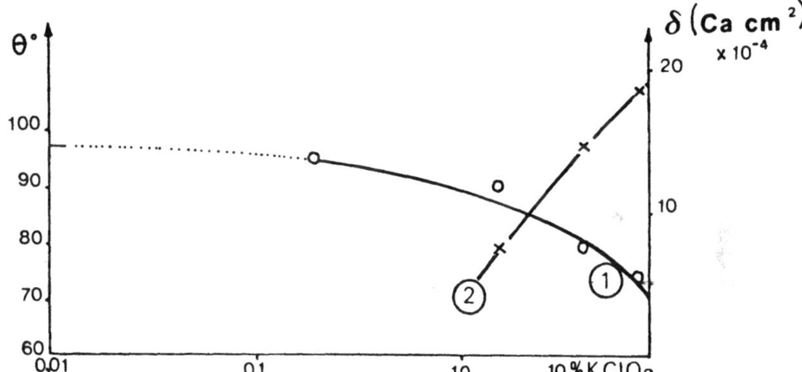

Fig. 2. Contact angle Θ and surface density δ of polar sites vs. the oxidizing mixture composition; 20 °C. 30 sec. immersion time

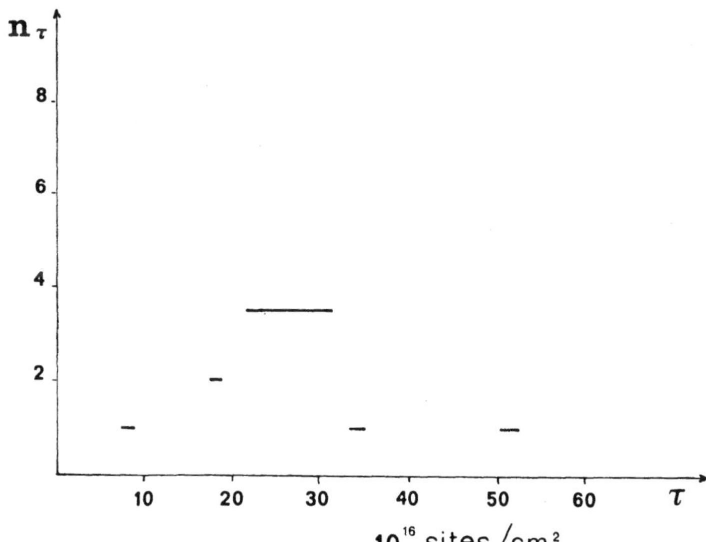

Fig. 3. Distribution curve of degree of grafting. τ = number of Ca adsorbing sites on PE. n_τ = number of samples corresponding to a value of τ. Time of grafting 4 hours

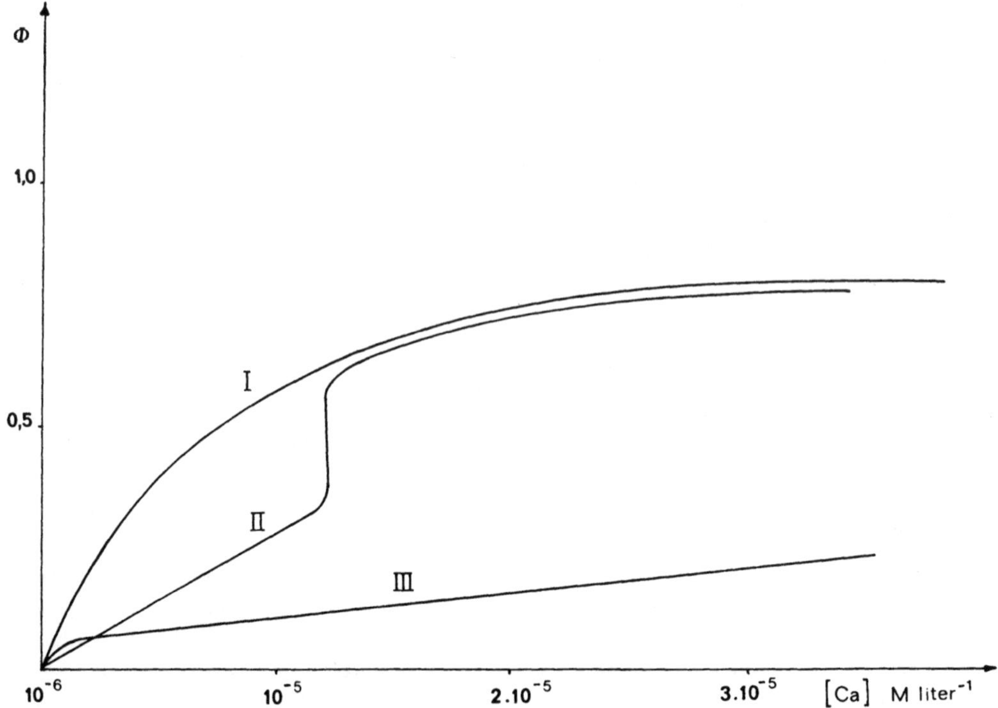

Fig. 4. Isotherms of adsorption of Ca on grafted PE. Φ = fraction of sites occupied by Ca ions. Ca = concentration of free Ca ions in solution. I and III = Time of grafting: 2 hours. II = Time of grafting 4 hours — pH = 7.6. I and II: concentration of Na$^+$ ions: 6.6×10^{-3} (N) III: concentration of Na$^+$ ions: 0.3 (N). Room temperature

Fig. 4 represents two isotherms corresponding respectively to a low degree of grafting (I) or a high degree of grafting (II). A "phase like" transition may be noted on the curve II when the binding of Ca ions exceeds 0.35 which represents about 50% of the maximum Ca binding at pH = 7.5. This transition does not occur at pH = 6 or with the lesser grafted samples (curve I). It may depend on the morphology of the grafted PMA, different for the low or for the high degrees of grafting.

The competition between the ions Ca and Na for the carboxyls of the PMA is demonstrated by the curves I and III of fig. 4. It appears that the binding of Ca ions is in fact a Na/Ca ion exchange which occurs whenever the ionic strength of the aqueous phase is increased by addition of the supporting electrolyte NaCl. The process has been discussed in reference (8).

2. Wettability of modified PE

The work of adhesion has been calculated from the contact angles by the classical expression:

$$W = \gamma_L (\cos \theta + 1) + \pi^s \qquad [1]$$

assuming that the spreading pressure π^s on the solid was negligible.

a) Grafted PE

The results are reported in table 2. They were obtained for a value of $\gamma_L = 72.2$ dyne/cm.

They demonstrate the specific effect of the nature of the monovalent counterion on the wettability of grafted PE. The effect is opposite to that one can expect if hydration energy were determining the wettability (see a). Furthermore an increase in the amount of Ca ions increases the wettability. These effects have not been explained yet.

The work of adhesion for the organic liquid: decalin + methylene iodide increases with the amount of grafting of PE and even more in the presence of Ca ions on the surface (20) (table 3).

b) Oxidized PE

Table 4 reproduces the results obtained at 23 °C for the system oxidized PE=water. In Fig. 5 the work of adhesion is plotted vs. δ, the surface density of the carbonyl groups measured by Ca adsorption. A saturation effect may be noted.

Table 1. Surface Grafting of Polyethylene

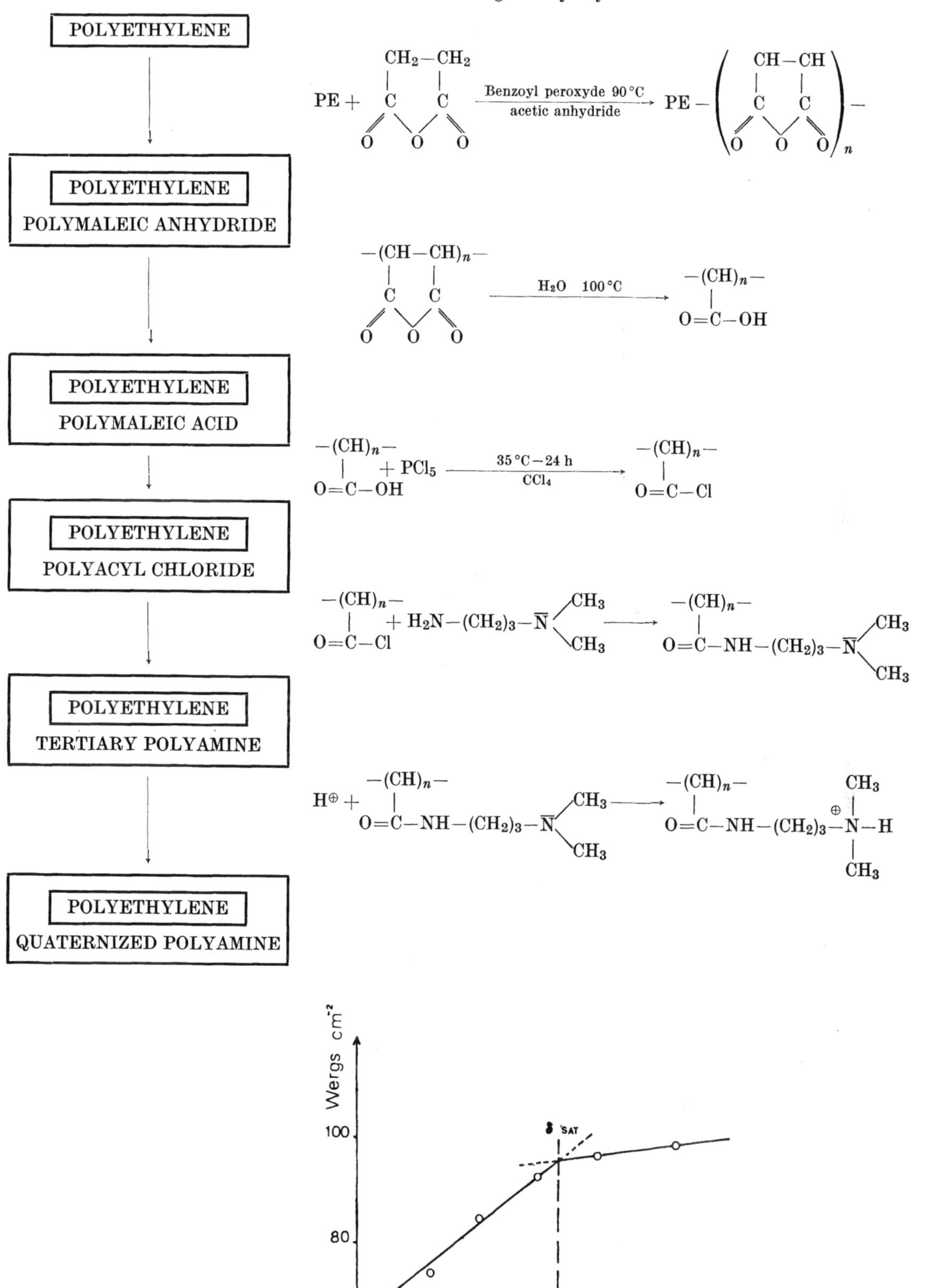

Fig. 5. Work of adhesion with water vs. surface density δ of polar groups on PE. δ_{sat} = saturation density

Table 2. Work of adhesion of PMA grafted PE. Specific effect of counterions

a) Constant ionic composition of the surface $\Phi = 0$

ion	Li	Na	K
$r_{(A)}$	0.6	0.95	1.33
$\Theta_A°$	68	59	57
W (ergs cm^{-2})	100	110	112

b) Varying surface ionic composition

Φ	0.28	0.58	0.65	0.74	0.83
$\Theta_A°$	59	58	58	53	43
W (ergs cm^{-2})	109	110	110	115	125

$\Theta_A°$ = advancing contact angle. Liquid: phosphate buffer. pH = 7.5; (Na) = 6.6×10^{-3} N. $\gamma_L = 73$ dyne cm^{-1}.

Grafting time $2^{1}/_{4}$ hours — r = unhydrated ionic radius — Φ = fraction of carboxyls occupied by Ca ions.

The effect of heating on the solid liquid adhesion is shown in fig. 6. The surface composition of the oxidized sample ($14,4 \times 10^{14}$ polar groups/cm^2) was the same for the three liquids used: water (1), ethylene glycol (2) and polyethylene glycol 200 (3).

Heating and cooling cycles have been performed for two more liquids: glycerol and formamide (not shown in the figure). The contact angle increases irreversibly when the temperature exceeds 80—85 °C. Subsequent heating and cooling runs do not display this irreversible aspect. The unoxidized PE does not show this irreversible behaviour. Fig. 7 represents the variation of the surface tension of the three liquids with temperature. No discontinuity may be noted on any one.

IV. Model of the interface and interpretation of results

This chapter will deal with the oxidized PE only.

Fig. 8 represents the model of PE-liquid interface. It has been imagined according to the conclusions of a previous work (9) and to a recent one (16). It represents the (010) face of cristalline PE, after complete oxidation. This face contains two methylene groups before oxidation. It is assumed that one of two methylene groups which may have been changed into carbonyl is exterior to the surface. Then the average area per carbonyl equals approximately 20 Å^2, as it has been assumed before (9, 22).

During the oxidation process the chemical attack may proceed beyond the (010) plane and

Table 3. Variation of work of adhesion of grafted PE with the amount of grafting

	$\delta_{Ca} \times 10^{-16}$ ion cm^{-2}	19.5	21.3	29.5	27.8	34.8	59
1.	$\Theta_A°$ (degrees)[a]	51		51			45
2.	$\Theta_A°$ (degrees)[b]		47		44	42	
3.	W (ergs cm^{-2})[a]	78		78			82
4.	W (ergs cm^{-2})[b]		81		82.5	84	

δ_{Ca} = surface density of Ca ions; $\Theta°$ advancing contact angles. Liquid: CH_2I_2 + Decalin (17% V/V). $\gamma_L = 48$ dyne/cm; 23 °C $\pm$ 1. [a]) Ca containing samples. [b]) Ca depleted samples by treatment with HCl 2 M.

Table 4. Contact angles, works of adhesion for oxidized PE-H$_2$O
Concentration of KClO$_3$ in the oxidizing mixture % (w/w): %

%	0	0.4	2.01	4.80	9.1	13	17
Θ_A (degrees)	96	94	88	80	74	70	68
W (ergs cm^{-2})	65	67	75	85.5	92	97	99

$\gamma_{H_2O} = 72.2$ dyne/cm.

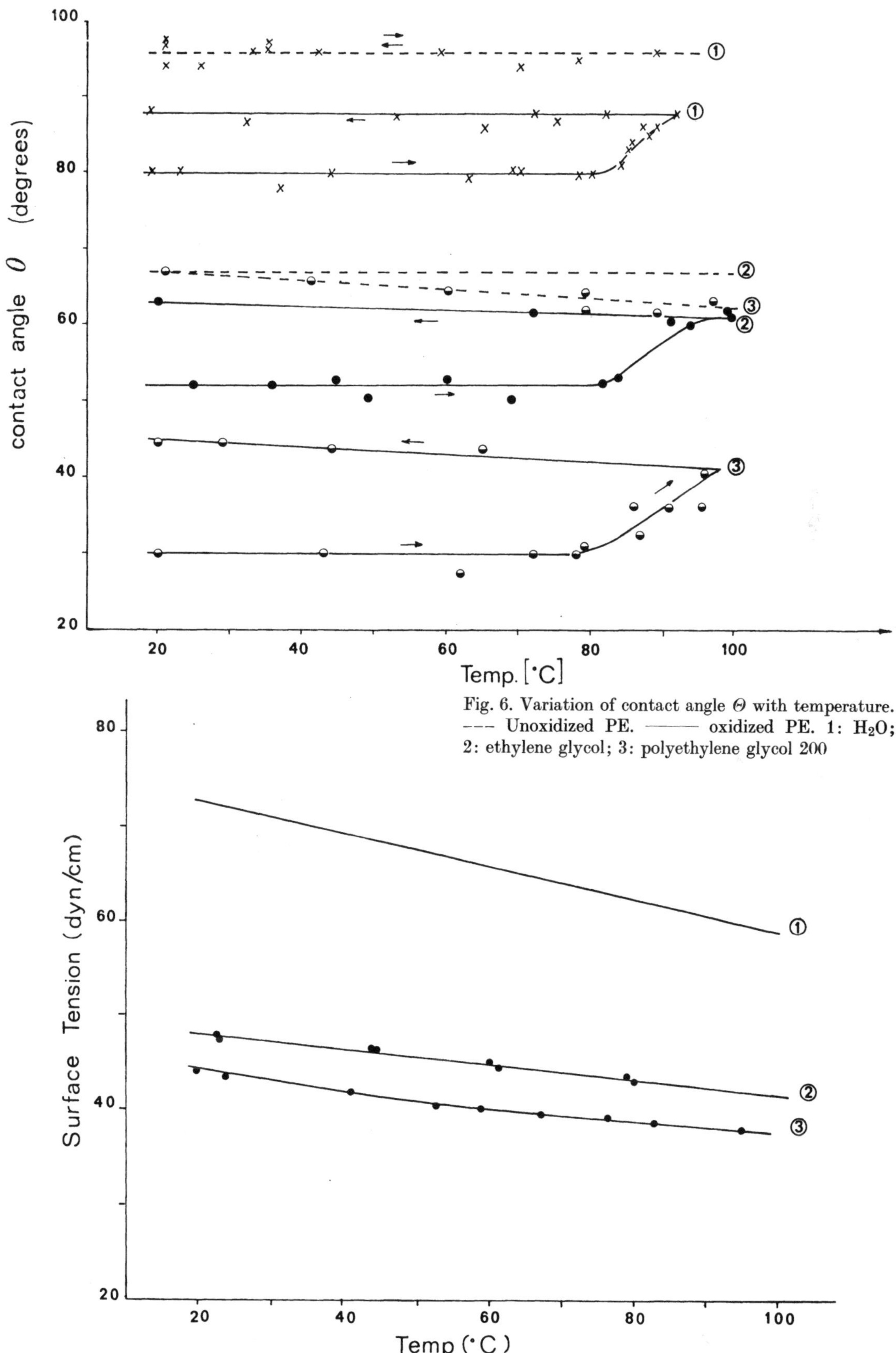

Fig. 6. Variation of contact angle Θ with temperature. --- Unoxidized PE. ——— oxidized PE. 1: H_2O; 2: ethylene glycol; 3: polyethylene glycol 200

Fig. 7. Variation of surface tension with temperature. 1: water; 2: ethylene glycol; 3: polyethylene glycol 200

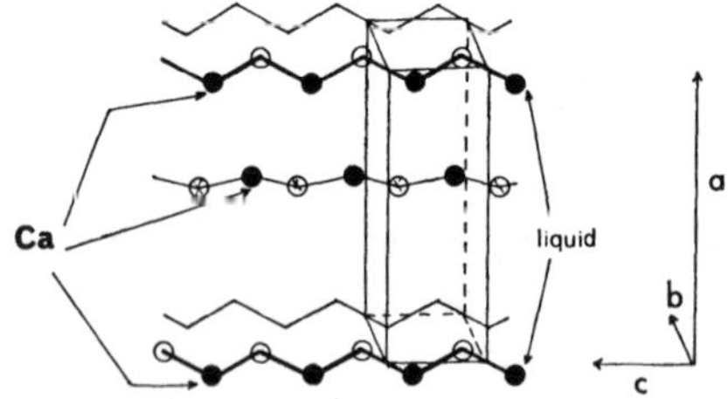

Fig. 8. Unit cell of polyethylene crystal. Ca: adsorbed ion on oxidized site

Ca ions penetrate below the surface. The molecules of the liquid do not follow. The saturation limit is only attained when the outermost carbonyls are bound to the liquid molecules.

a) Wettability of PE at room temperature

Let ω be the average area per external carbonyl and φ a degree of molecular roughness which allows for the oxidation of PE below the plane (010). Then δ/φ is equal to the external carbonyl group surface density and the work of adhesion is equal to

$$W = W^0 + \frac{\delta\,\omega}{\varphi}\,(W' - W^0)\ \text{erg cm}^{-2} \qquad [2]$$

where W^0, W' are respectively the works of adhesion of the unoxidized PE and of the completely oxidized one. The eq. [2] is valid for $\delta < \delta_{\text{sat}}$ (see fig. 4) where δ_{sat} is equal to 21×10^{14} groups cm^{-2}. Let $\omega = 20$ Å^2 group^{-1}.

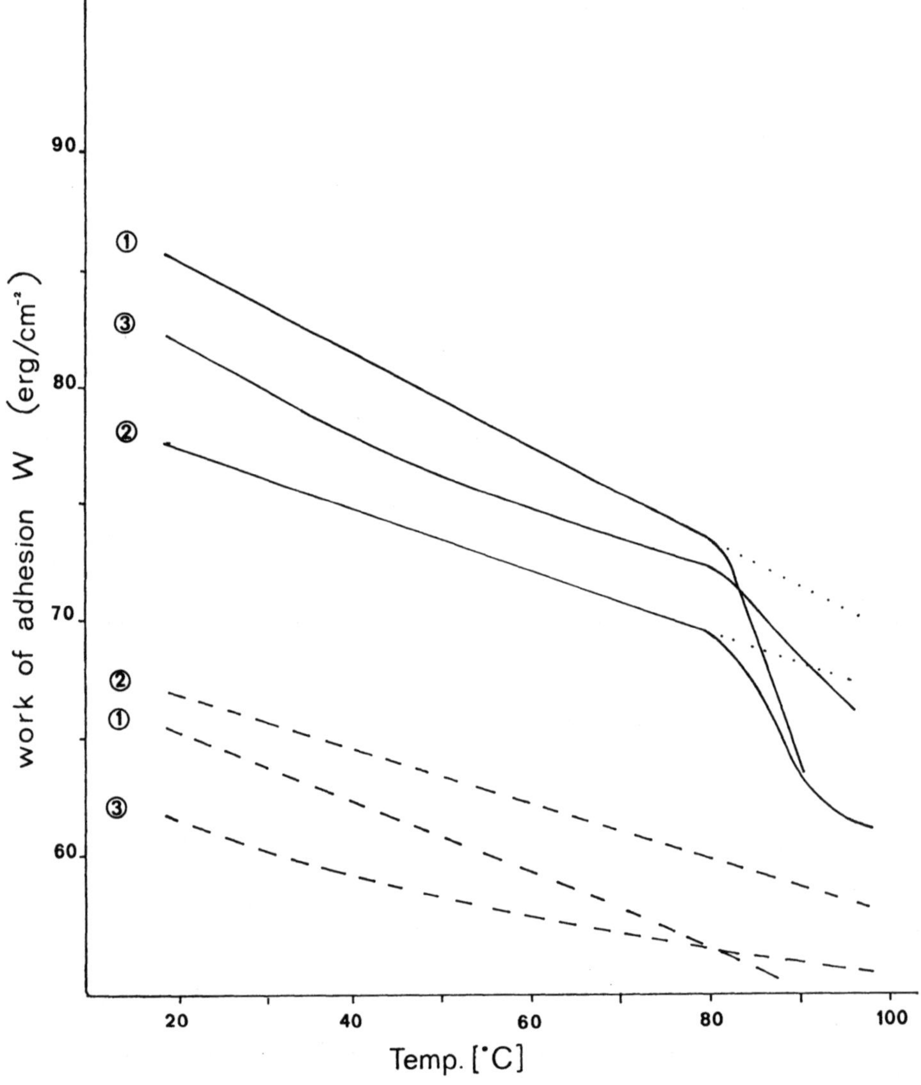

Fig. 9. Variation of the work of adhesion W with temperature. 1: water; 2: ethylene glycol; 3: polyethylene glycol 200

Then for $\delta = \delta_{sat}$, $W = W'$ and from [2] it is obtained that $\varphi = (\delta_{sat}\, \omega) = 4.2$. This means that one out of four surface carbonyl groups interacts with the liquid. From the slope of the steep line of fig. 4, and ω the dipole-dipole interaction energy ε_{dd} is calculated. ε_{dd} is equal to 18×10^{-14} erg group^{-1}, $4.7\ kT$ group^{-1} or 2.82 Kcal/mole^{-1}.

In reference (9) the theoretical value of ε_{dd} has been calculated assuming that

$$W' - W^0 = \frac{\varepsilon_{dd}}{\omega} = \frac{2}{\omega} \times \frac{\mu_{H_2O} \times \mu_{C=O}}{a^6} \qquad [3]$$

where ε_{dd} is the dipole-dipole interaction of a water molecule and a carbonyl group, the distance a between them being equal to 3.55 Å and $\mu_{H_2O} = 1.87$ D and $\mu_{CO} = 2.7$ D. It has been found $\varepsilon_{dd} = 22.6 \times 10^{-14}$ erg group^{-1}. The agreement between these two values is encouraging as it demonstrates that all the outermost carbonyl groups point in the direction of the polar liquid. This result may be explained by the

fact that the oxidation of PE takes place in an aqueous medium which favours this orientation. Furthermore PE being a solid this orientation corresponds to a frozen state of the polymers.

b) Effect of temperature on the wettability of PE

The irreversible decrease in work of solid liquid adhesion accompanying the melting of PE has been related to the beginning of molecular rotation at the surface (15). This process carries the carbonyl groups below the surface, beyond the range of the interaction forces with the liquid polar molecules.

The curves of fig. 9 represent the variation of the work of adhesion W with temperature, obtained from the values of Fig. 6 and the eq. [1]; the straight segment are parallel to the interrupted lines obtained for the unoxidized PE. It is assumed (16) that the linear segments correspond to the frozen, constant surface composition of PE, the decrease of the adhesion above 80—85 °C being due to the decrease in the

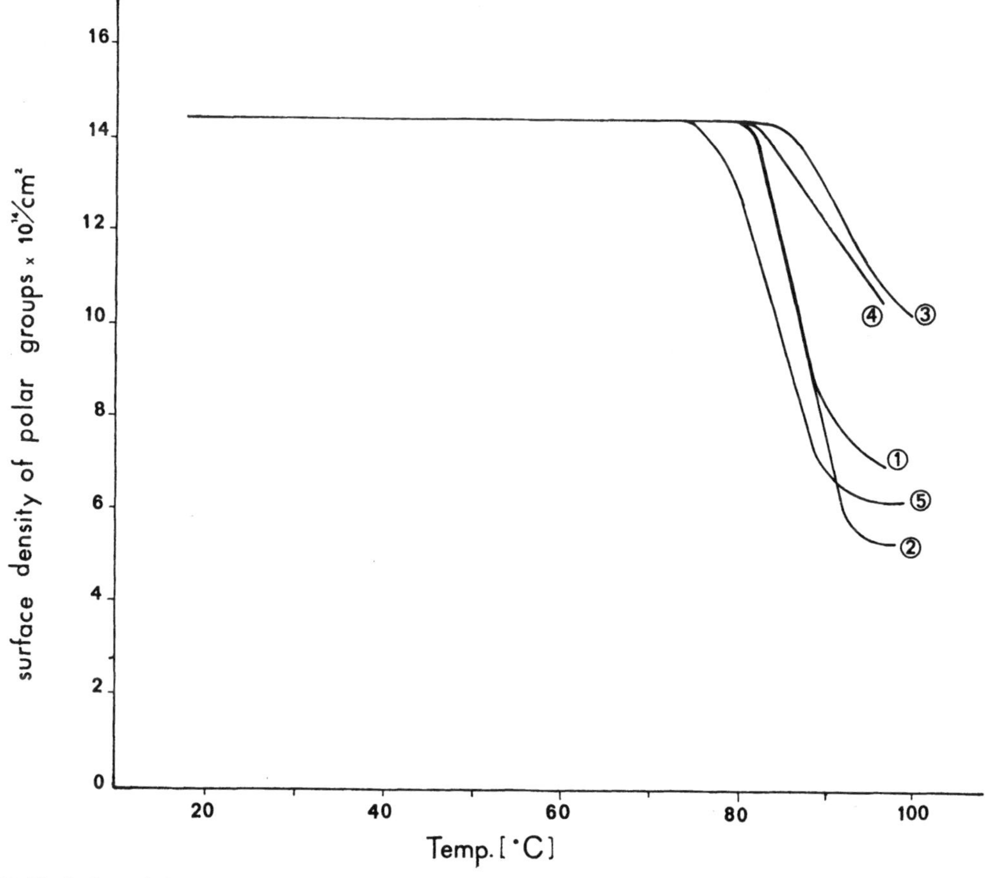

Fig. 10. Variation of the surface density of polar groups with temperature. 1: water; 2: polyethylene glycol; 3: glycerol; 4: polyethylene glycol 200; 5: formamide

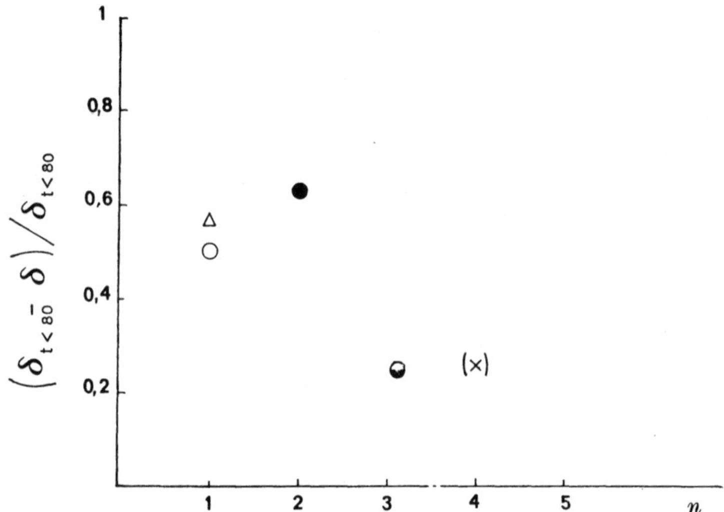

Fig. 11. Degree of over-turn of the superficial polyethylene chains vs. number n of hydrogen bonding functions of one molecule. ○: water; ●: ethylene glycol; ◐: glycerol; (×): polyethylene glycol 200; △: formamide. $\delta t < 80$: surface density of oxidized sites at $t < 80\,°C$

value of δ in eq. [2]. Using the results for W shown in fig. 9, and the eq. [2] the variation of δ with the temperature has been deduced as follows:

$$\delta(t) = 14.4 \times 10^{14}\,\frac{W_t - W^0}{W' - W^0} \qquad [4]$$

where 14.4×10^{14} is the low temperature, initial surface density of the carbonyls and the dispersion term W^0 is independent of t.

Fig. 10 represents the results for the five liquids studied. They divide into two groups: polyethylene glycol 200 and glycerol, on one side, and water, glycol and formamide, on the other side. Also, the decrease in δ starts at temperatures increasing in the order formamide $<$ water $\cong$ ethylene glycol $<$ polyethylene glycol 200 $<$ glycerol. Furthermore at high temperatures, the decrease in δ seems to tend to a limit specific to each liquid.

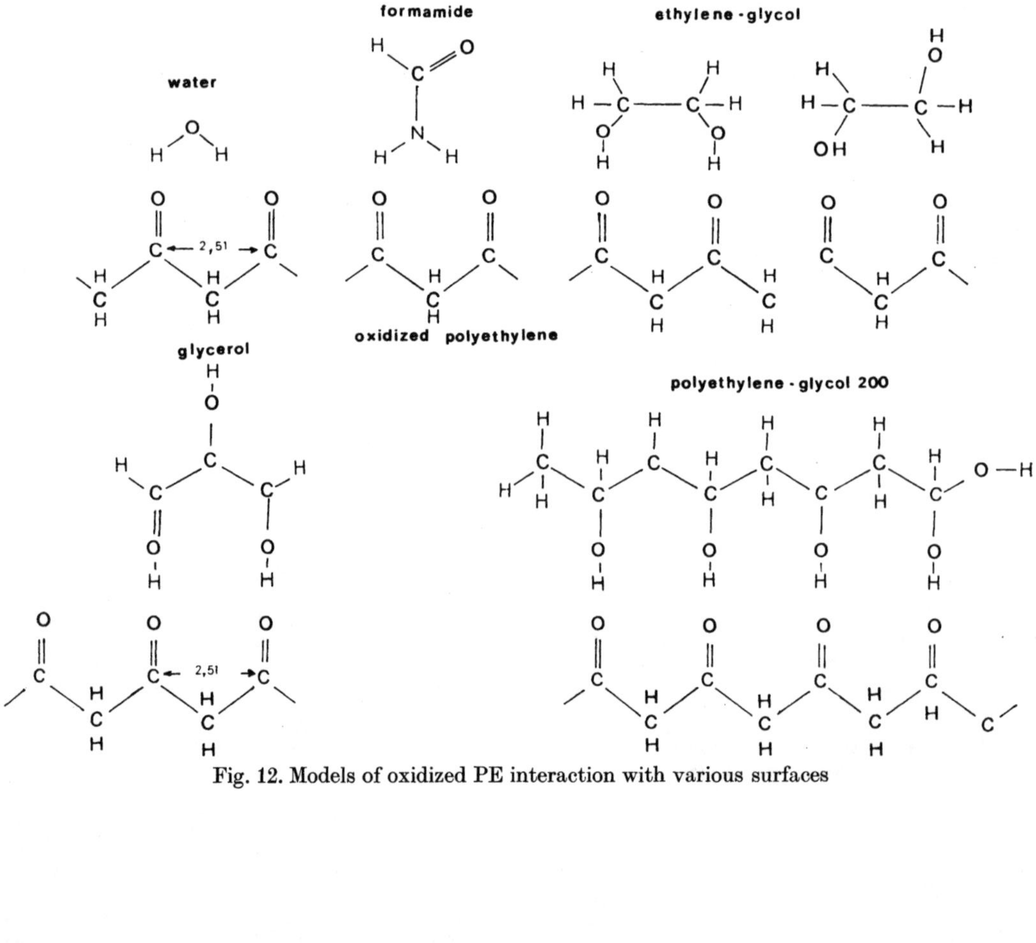

Fig. 12. Models of oxidized PE interaction with various surfaces

It may be noted that all the liquids utilized are hydrogen bonding. The number n of hydrogen bonds per molecule increases from one for water to a maximum of four for polyethylene glycol 200 which contains four hydroxyls. In fig. 11 the percentage of the outermost carbonyl groups left at high temperature is plotted vs. n. The value corresponding to polyethylene glycol 200 is presumably too high as this substance has not been studied above 95 °C. The decrease in the number of adhesive bonds with the small molecules is about 50% and about 25% only when the molecules of the liquid are large and carry several hydrogen bonding groups. Ethylene glycol seems to be a particular case.

Fig. 12 illustrates explanation of the results shown in fig. 11. The protection of the adhesive bonds against their destruction by thermal motions may be obtained by bridging together several sites of the macromolecules of the polymer to the large molecules of the liquid. This may be done successfully when the structure of liquid molecules allows a distance between two hydroxyls equal to 2.51 Å at least, without additional tension. Such is the structure trans of glycol and polyoxyethylene 200, but not of ethylene glycol. According to our results (fig. 11), this last substance belongs to the class of the small molecules, showing that, on the average, it has the transconformation at the oxidized PE surface.

Discussion and concluding remarks

The highly idealistic model of cristalline PE may be criticized. While being aware of the fact that the model of fig. 8 is leading to the satisfactory average value of 20 Å² for one carbonyl on the surface, we note that the PE we have studied is obtained by extrusion at an elevated temperature under conditions where stretching of the macromolecules of the polymer may produce orientation of the macromolecules parallel to the surface of the film and partial cristallization.

When grafting, or oxidation of the polymer is performed the increase in adhesion to polar liquids increases with the degree of the surface chemical treatment.

The PE surfaces grafted with PMA behave like "two-dimensional" cation exchangers. Their wettability depends on the nature of the cations fixed. But on account of their surface roughness

the interpretation of the results is difficult. Studies are carried out on the effect of temperature on their wettability by hydrogen donor liquids.

The results of the studies carried out for oxidized PE are understood easier.

Each carbonyl formed on the outermost plane of PE by oxidation may form a hydrogen bond with a hydrogen donor group. The increase in the work of adhesion provided by these hydrogen bonds corresponds, on the average, to 2.8—2.9 Kcal mole^{-1}. This value is smaller than that for a $\rangle C{=}O \cdots H{-}O{-}$ bond in the acetic acid dimer, but comparable to the value 2.5 Kcal for the $\rangle C{=}O \cdots H{-}$ bond in chloroform acetone mixtures (24). It is known that the energy of the hydrogen bond varies with its length.

As far as movements of macromolecules at interfaces are concerned, the effect of the binding of the macromolecule by two bonds to an interface, even with a liquid, is interesting. It is known that macromolecules adsorbed at a solid liquid interface unfold to lower their free energy. In our case, the lowering of the free energy is presumably used to hinder the tendency of the stretched surface macromolecules to coil up and transfer their polar groups away from the interface.

This conclusion may have biological implications of the macromolecule in a membrane protein.

Thus if for some reason, this protein undergoes a change in conformation, the adhesivity of the membrane may be modified. The extent of this last modification will vary according to the nature of the adherent and, in particular, to its possibility to establish molecular bonds with the membrane protein.

References

1) *Borisova, T. K., A. V. Kiselev, A. Ya. Korolev, V. L. Lygin,* and *L. N. Salomonova,* Kolloidn. Zh. **28,** 792 (1966).
2) *Porejko, S., W. Gabara,* and *J. Kulesza,* J. Polymer Sci. **A 1, 5,** 1563 (1967).
3) *Kaelble, D. H.,* Phys. Chem. of Adhesion (New York 1971).
4) *Fowkes, F. M.,* Ind. Eng. Chem. **56** ,40 (1964).
5) *Good, R. J.* and *L. A. Girifalco,* J. Phys. Chem. **44,** 561 (1960).
6) *Baszkin, A.* and *L. Ter-Minassian-Saraga,* C. R. Acad. Sc. Paris, Serie C, **268,** 315 (1969).

7) *Baszkin, A.* and *L. Ter-Minassian-Saraga*, J. Polymer Sci., Part C, **34**, 243 (1971).

8) *Baszkin, A.* and *L. Ter-Minassian-Saraga*, J. Colloid Interface Sci. **43**, 478 (1973).

9) *Baszkin, A.* and *L. Ter-Minassian*, J. Colloid Interface Sci. **43**, 190 (1973).

10) *Baszkin, A.* and *L. Ter-Minassian-Saraga*, J. Colloid Interface Sci. **43**, 473 (1973).

11) *Lee, L. H.*, in: Interaction of Liquids and Solid Substrates. Adv. Chem. Series 87, P. 106 (1969).

12) *Rehbinder, P. A., M. Lipetz, M. Rimskaya,* and *A. Taubmann*, Kolloidn. Zh. **65**, 268 (1933).

13) *Ter-Minassian-Saraga, L.*, J. Colloid Interface Sci. **22**, 311 (1966).

14) *Blais, P., D. J. Carlsson, G. W. Csullog,* and *D. M. Wiles*, J. Colloid Interface Sci. **47**, 636 (1974).

15) *Baszkin, A.* and *L. Ter-Minassian-Saraga*, Polymer **15**, 759 (1974).

16) *Baszkin, A., N. Nishino,* and *L. Ter-Minassian-Saraga*, Submitted paper.

17) *Ter-Minassian-Saraga, L.*, C. R. Acad. Sc. Paris, Groupe 7, **260**, 5770 (1965).

18) *Deyme, M.* and *A. Baszkin*, C. R. Acad. Sc. Paris, Série C., **278** (1974).

19) *Baszkin, A.*, Thesis, Figures III-2 and III-3 (Paris 1973).

20) *Martinet, J. M.*, Private communication.

21) Handbook of Chemistry and Physics, 44th ed., p. 2239 (1963).

22) *Fowkes, F. M.*, J. Colloid Interface Sci. **28**, 493 (1968).

23) *Bunn, C. W.*, Trans. Faraday Soc. **35**, 482 (1939).

24) *Pauling, L.*, Nature of the Chemical Bond, 3rd ed. (Ithaca 1960).

Authors' address:

A. Baszkin, M. Deyme, M. Nishino,
and *L. Ter-Minassian-Saraga*
Equipe de Recherche C.N.R.S. associée
á l'Université Paris V
U.E.R. Biomédicales 81, Paris (France)

Für die Schriftleitung verantwortlich: Für Originalarbeiten Prof. Dr. F. H. Müller, 3550 Marbach b. Marburg/L.
und Prof. Dr. Armin Weiss, 8000 München 2
Dr. Dietrich Steinkopff Verlag GmbH & Co. KG, Postfach 11 10 08, Saalbaustraße 12, 6100 Darmstadt 11
Herstellung: Konrad Triltsch, Graphischer Betrieb, 8700 Würzburg

Mitarbeiter-Bedingungen · Note to Contributors

Originalbeiträge sind an die folgenden Herren zu senden:

Arbeiten aus dem Bereich der Polymerforschung an
Prof. Dr. F. H. Müller (Direktor des Univ.-Instituts für Polymere, Marburg/Lahn) Haselhecke 26, 3550 Marburg-Marbach;

Arbeiten aus dem Bereich der Kolloidchemie und Biochemie an
Prof. Dr. A. Weiss (Institut für Anorganische Chemie der Universität München) Meiserstraße 1, 8000 München 2;

Die Zeitschrift veröffentlicht nur angeforderte Originalbeiträge zu jeweils einem bestimmten Thema pro Band.

Manuskripte sollen in zweifacher Ausfertigung eingereicht werden. Ihr Eingang wird umgehend bestätigt. Ihr Inhalt muß unveröffentlicht sein. Die Verantwortung für den Inhalt liegt bei den Autoren. Publikationssprachen: Deutsch, Englisch oder Französisch. Jedem Manuskript ist eine Zusammenfassung in deutscher und englischer Sprache beizugeben. Die Typoskripte müssen einseitig und weitzeilig geschrieben sein. Abbildungen sind mit Legenden zu versehen und als klischierfähige Vorlagen einzureichen, wobei die Beschriftung auf einem transparenten Deckblatt anzubringen ist. Formeln bitte deutlich schreiben, insbesondere griechische Buchstaben und Indices! Die Zahl der Abbildungen und Tabellen ist auf das unbedingt Notwendige zu beschränken. Für Literaturangaben gelten die international üblichen Regeln. Die Literatur ist am Schluß der Arbeit zusammenzufassen. — Anstelle eines Honorars erhalten die Autoren insgesamt 75 Sonderdrucke kostenlos. weitere Exemplare auf ausdrücklichen Wunsch gegen Berechnung. — Kosten für nachträgliche Autorkorrekturen, soweit es sich um Textergänzungen in der Druckfahne handelt, werden dem Autor in Rechnung gestellt. — Ausführliche Sonderdrucke der geltenden Mitarbeiterbedingungen sind kostenlos beim Verlag erhältlich. — Nicht den Richtlinien entsprechende Manuskripte werden zurückgesandt.

Der Verlag erwirbt mit der Annahme des Manuskriptes das ausschließliche Recht der Vervielfältigung, gewerbsmäßigen Verbreitung, Übersetzung und Verwendung für fremdsprachige Ausgaben der in dieser Zeitschrift erscheinenden Beiträge. Gleichzeitig überträgt der Autor gemäß § 54 URG dem Verlag auch das Recht, die Herstellung von photomechanischen, xerographischen oder sonstigen Vervielfältigungen seines Beitrages oder eines Teils desselben nach Maßgabe des zwischen der Verwertungsgesellschaft Wissenschaft GmbH (ehemals Inkassostelle für urheberrechtliche Vervielfältigungsgebühren GmbH) und dem Bundesverband der Deutschen Industrie sowie anderen Verbänden abgeschlossenen Gesamtvertrages vom 15. 7. 1970 zu genehmigen. Diese Genehmigung bezieht sich auf die Herstellung von derartigen Vervielfältigungen in gewerblichen Unternehmen zum innerbetrieblichen Gebrauch. Das Abkommen sieht vor, daß 50 % des Reinerlöses zugunsten eines Urheberfonds verbucht werden. Die Weitergabe von Vervielfältigungen, gleichgültig, zu welchem Zwecke sie hergestellt wurden. ist verboten und als Urheberrechtsverletzung strafbar.

Die Wiedergabe von Gebrauchsnamen, Handelsnamen. Warenbezeichnungen usw. in dieser Zeitschrift berechtigt auch ohne besondere Kennzeichnung nicht zu der Annahme, daß solche Namen im Sinne der Warenzeichen- und Markenschutz-Gesetzgebung als frei zu betrachten wären und daher von jedermann benutzt werden dürften.

The authors are requested to submit their **manuscripts** to the following Editors:
Contributions on Polymer Science to

Contributions on Colloid Science and Biochemistry to

This journal will publish original contributions only on request by the editors covering the special scope of each volume.

Manuscripts should be submitted in duplicate and should contain original work as yet unpublished elsewhere. Their receipt will be acknowledged promptly. Authors are fully responsible for the contents of their contributions. Publications languages: English, French or German. Each manuscript should include a summary in English and German. All manuscripts should be double-spaced, typed on one side only. Illustrations and drawings should be made carefully, with India ink on white drawing paper, blue tracing linen or coordinate paper ruled in blue only. Lettering at the sides of graphs may be pencilled in and will be typeset. Legends must accompany the drawings, Formulas, symbols and Greek letters should be carefully made and annotated and subscripts and superscripts clearly shown. The number of figures and tables should be held to a minimum. The list of references should be written on a separate page. It is recommended that abbreviation of the titles of the Journals be made in conformity with Chemical Abstracts (see List of Periodicals, 1961).

Authors will receive 75 reprints of their contribution free of charge and may order an additional number at cost. Authors making elaborate alterations and additions in proof will be required to bear the costs thereof. More detailed instructions to the authors can be obtained from the publisher free of charge. Manuscripts which do not conform with the above guidelines will be returned to the authors.

By accepting the manuscripts the publisher acquires the sole right of reproducing, selling, translating and using it for foreign language editions. The author also gives the publisher the right of photostating, xerographing and otherwise reproducing the paper or part of it in accordance with § 54 German Copyright Law (URG) and the Agreement of the Verwertungsgesellschaft Wissenschaft GmbH (formerly Inkassostelle für urheberrechtliche Vervielfältigungsgebühren GmbH) and the Bundesverband der Deutschen Industrie and other similar institutions of July 15, 1970, respectively. This permission includes reproduction by an industrial organization for internal use only. The Agreement cited above provides that 50 % of the net profit is to be paid into the account of a Copyright Fund. The distribution of any reproduced material to other persons or institution is prohibited and will be prosecuted as a violation of the copyright laws.

The reproduction of brand names, trade names, trade marks etc. in this journal should not be interpreted to mean that such names are not covered by the Trademark and Tradename laws, and that they can be used freely.

Geschäftliche Bedingungen · Note to Subscribers

Erscheinungsweise:
Zwanglos nach Bedarf in Bänden verschiedenen Umfangs.

Frequency of Publication:
Irregularly in volumes of different size.

Bezugspreis dieses Bandes:
DM 70,– plus Porto.
Bezieher der Kolloid-Zeitschrift & Zeitschrift für Polymere erhalten den Band automatisch im Rahmen ihres Abonnements mit 20 % Nachlaß.
Die Zeitschrift wird automatisch zur Fortsetzung weitergeliefert, sofern nicht vier Wochen vor Jahresende eine Abbestellung vorliegt.

Subscription rate of this volume:
DM 70,– plus postage.
Subscribers to Kolloid-Zeitschrift & Zeitschrift für Polymere will receive this volume additionally with a 20 % discount.
The subscription will be extended automatically for unless there is a cancellation received four weeks before the end of each year.

Photokopier-Wertmarken:
Für jedes Photokopierblatt eines Beitrages oder Beitragsteiles aus dieser Zeitschrift ist eine Wertmarke von DM –,40 zu verwenden, erhältlich bei der Inkassostelle für urheberrechtliche Vervielfältigungsgebühren GmbH., 6000 Frankfurt a. M. 1.

Photostat-Stamps:
Each photostat-sheet of an article or part of an article published in this journal must show a stamp of DM –,40, which may be obtained by the Inkassostelle für urheberrechtliche Vervielfältigungsgebühren GmbH., 6000 Frankfurt a. M. 1.

Verlag, Copyright und Anzeigenverwaltung:
Dr. Dietrich Steinkopff Verlag GmbH & Co. KG, Postfach 11 10 08, 6100 Darmstadt 11, Telefon (Phone) (06151) 2 65 38/9 - Postscheckkonto (Postal Account) Frankfurt a. M. 956 97 - Bank (Bankers): Deutsche Bank Darmstadt No. 026/0117. Foreign subscribers are advised to pay by cheque.

Publisher, Copyright and Advertising Manager:

Titel-Abkürzung:

Abbreviation of Title:
Progr. Colloid & Polymer Sci.

Made in the USA
Monee, IL
07 July 2026

56549920R00066